RICHARD A. LARSON, Editor

BIOHAZARDS
of
Drinking Water
Treatment

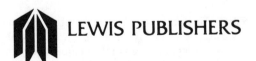

LEWIS PUBLISHERS

Library of Congress Cataloging-in-Publication Data

Biohazards of drinking water treatment.

  Bibliography: p.
  Includes index.
  1. Drinking water—Purification. 2. Drinking
water—Contamination. 3. Health risk assessment.
I. Larson, Richard A.
TD433.B56 1988   363.6'1   88-13185
ISBN 0-87371-110-6

LEWIS PUBLISHERS, INC.
121 South Main Street, Chelsea, Michigan 48118

PRINTED IN THE UNITED STATES OF AMERICA

Preface

The chapters in this book were first presented as papers at a symposium of the Environmental Chemistry Division of the American Chemical Society at its 194th national meeting, held in New Orleans, Louisiana, in September 1987. The symposium called attention to the many problems that remain in trying to provide drinking water of good quality to the people of industrialized societies, in spite of the formidable advances that have been made over the past century.

Although evidence of the use of some crude water purification techniques was recorded in manuscripts dating back to 2000 B.C., widespread recognition that drinking water needed to be protected from sources of bacterial contamination such as sewage did not occur until the nineteenth century. The first efforts to deliberately treat drinking water on a large scale were slow sand filtration processes, first practiced in England in 1829 and in the United States in 1870. By the beginning of World War I, approximately 30% of the U.S. population was drinking filtered water. Filtration, although widespread, is still not universal; about a third of all treatment facilities in the United States do not filter their source water.

In 1908, an even more important practice in water treatment had been established—drinking water chlorination. The dramatic decreases in waterborne infectious diseases that followed the introduction of disinfection by chlorine led to a rapid and almost uncritical acceptance of this technique during the first half of this century. Disinfection (usually with chlorine) is now almost universally practiced in the United States. Yet, virtually the only problem related to water chlorination that attracted any chemical attention at all for more than 50 years was the formation of taste-and-odor causing chlorophenols. However, in the early '70s, the first reports of trihalomethanes (haloforms) in public drinking water supplies appeared, and the realization that these substances were produced by water chlorination followed very soon afterwards. Because of evidence that these compounds might be implicated in some forms of cancer, public concern was considerable. As a result of these findings and concerns, a significant reappraisal of disinfection practices began throughout the world and is continuing. Trihalomethane levels in finished drinking water were limited by the U.S. EPA to less than 100 parts per billion, and most other developed countries have adopted similar regulations. In Europe, there has been a tendency not to chlorinate and to rely on alternative methods to maintain drinking water quality, whereas in North America chlorine is still very widely used, although application practices have generally changed somewhat in order to minimize trihalomethane concentrations.

Waterborne disease outbreaks in the United States began increasing in the '70s, after 20 years of very low levels. Many of these episodes have involved the occurrence of pathogenic protozoa such as *Giardia* that are able to persist in periods of low disinfectant concentrations. Although most of these incidents have occurred in small communities, the trend is sobering and serves to point up the necessity for continuing reevaluation of public policies relating to drinking water treatment.

The opening section of this book is devoted to an overview of public health risks related to drinking water treatment practices and the perception of these risks by users. Distinct potential hazards are associated with the use of chlorine, chlorine dioxide, and ozone—different classes of oxidation products from dissolved organic matter, different degrees of disinfection efficacy, different periods of effectiveness in the distribution system—and all of these problems need to be addressed and evaluated by water authorities, especially in the light of public response to information about these hazards. As new kinds of complex and potentially risky problems related to drinking water are discovered (high nitrate levels and new classes of potential pathogens, for example) the question of how to present these findings becomes more and more critical. The plant manager needs not only to be knowledgeable in biology and chemistry, but also must know something about basic psychological principles.

The second section consists of chapters on chlorinated ethanes—volatile organic compounds that are becoming more and more frequently detected in contaminated source waters, especially groundwater. The sources of these chemicals and their fates during water treatment are not fully understood, but the potential for long-term health effects seems significant. Furthermore, it is not sufficient merely to measure the concentrations of these chemicals, since it seems clear that they can undergo transformations into other products with different toxicologic potentials.

In the third section of the book, the question of pathogenic organisms in treated drinking water is examined. Viruses are particularly difficult microorganisms to monitor, and past evidence has indicated that some types are becoming more resistant to common methods of disinfection. Work on identifying waterborne viruses and their abilities to survive the rigors of drinking water treatment is thus becoming more crucial. Again, although certain types of bacteria, such as coliforms, are routinely monitored by almost all water plants, further work to detect other potentially pathogenic classes should also be carried out on occasion, even if only to show that their numbers are too low to represent a significant infective dose. Finally, although there have been very few studies of fungi in drinking water systems, it is clear that they do occur there, and additional work is required to assess the importance of the types identified, which include some known or suspected human pathogens.

The fourth section of the book examines some of the chemical reactions between aqueous chlorine and common constituents of natural waters such as amino acids and humic materials, and delineates some of the efforts that have gone into identifying particularly active mutagens found in drinking water concentrates. Although great strides have been made in the past decade in our understanding of the chemistry of the reactions of aqueous chlorine with organic compounds, much remains to be learned about the biological significance of the types of products that are formed.

Because of the recognized health risks of exposure to water chlorination products, interest in the use of ozone as an alternative disinfectant has become general. More and more water authorities are considering the use of ozone either as the sole disinfectant or in combination with others, such as chlorine. However, as the chapters in the fifth section of the book point out, we really know very little about the products formed by ozonolysis of aquatic organic matter and their potential effects. In particular, it appears that organic oxidants and polymers may form under some conditions; these have not been carefully analyzed in most previous studies. It would seem to be imperative that careful studies of ozonation products and their biological effects should continue before ozone is adopted widely. In particular, some of the aliphatic aldehydes produced by ozone should be studied painstakingly, given recent reports of their mutagenicity.

Granular activated carbon (GAC) is the subject of the sixth section. This material has become widely used in recent years to remove hydrophobic contaminants from source water and treated waters. Under some conditions, GAC can efficiently remove toxicants from chlorinated water. Also, recent studies have shown that GAC is not an inert support, but is converted by oxidizing agents such as disinfectants to a free-radical oxidant capable of converting compounds like phenols and anilines to many new products. Furthermore, the high surface area of GAC is ideal for permitting the growth of large numbers of bacteria. The sum total of all the positive and negative aspects of the use of GAC in water treatment cannot yet be evaluated, but it seems clear that it is not a panacea.

Finally, the last section of the book, entitled "Recent Developments," includes several chapters on a variety of topics that represent advanced thinking about fundamental aspects of water treatment. Should we synthesize totally new disinfectants? Should we use the new techniques of molecular biology to probe for trace levels of viral contamination? And, should we reconsider some fundamental concepts of water treatment so that we do not try to kill organisms that are present, but to prevent their growth in the first place?

It is hoped that not only these last chapters, but all of the chapters in the book, will stimulate thought about our basic attitudes toward drinking water, and help to bring about the ideal of the safest drinking water for the greatest number. It needs to be pointed out that on a global basis, clean drinking water can by no means be taken for granted. The majority of the world's people do not have access to clean water, and as a result some 80% of all disease is due, ultimately, to improperly sanitized food and water. It has been estimated that up to 25,000 deaths of infants and small children each day occur because of diarrhea, often spread by impure water. However, this huge and deplorable problem is, unfortunately, beyond the purview of the authors of the present book. It is to be hoped that governments and scientific societies throughout the world will be able to bring attention and funding to bear in order to help alleviate some of the incalculable misery of the world's poor through universal provision of hygienic water supplies.

Contents

SECTION IV
CHLORINATION PATHWAYS AND BY-PRODUCTS

SECTION V
OZONATION BY-PRODUCTS

SECTION VI
GRANULAR ACTIVATED CARBON

SECTION VII
RECENT DEVELOPMENTS

SECTION I

Overview

CHAPTER 1

Health Risk in Relation to Drinking Water Treatment

Henk J. Kool, Stichting Waterlaboratorium Oost, Doetinchem, The Netherlands

INTRODUCTION

In The Netherlands about 70% of drinking water is produced from groundwater.[1] The remaining 30% has to be prepared from surface waters, mainly from the rivers Rhine and Meuse. These rivers are polluted with inorganics, organics, and pathogenic microorganisms.[2-5]

In order to be certain that the waterworks produce safe drinking water, treatment processes have to be applied. An overview of the most frequently used treatments in drinking water supply in The Netherlands is shown in Table 1.

It shows that the majority of the distributed drinking water (about 740 $\times$ 10^6 m^3 yearly derived from groundwater) does not undergo a chlorine treatment. Over many decades of this "treatment" practice, no significant problems have ever occurred with respect to the hygienic safety of drinking water prepared from groundwater. The use of a chlorination step, or other oxidation steps like ozone and chlorine dioxide, is related to drinking water derived from surface water.

Chlorination in drinking water supply generally is applied in The Netherlands as transport, or safety postchlorination. Application of ozone in this respect is mainly applied to improve the water quality (for instance, taste and odor), but at the same time it also serves a hygienic purpose.

In this chapter, effects of several oxidation treatments (viz., chlorine, chlorine dioxide, and ozone) on the water quality will be summarized and related to possible toxic effect caused by organics. In addition, experience of the Amsterdam Waterworks, which recently has stopped its safety chlorination, will be discussed briefly. Finally, problems in drinking water quality

Table 1.   Drinking Water Treatment in Water Supplies in The Netherlands

Raw Water Source		
Surface Water I	**Surface Water II**	**Groundwater**
Coagulation	Coagulation	Aeration
Open storage	Rapid sand filtration	Rapid sand filtration
Transport chlorination[a]	Transport chlorination[a]	
Ozonation	Artificial recharge in dunes	
Activated carbon	Aeration	
Coagulation	Activated carbon	
Rapid sand filtration	Rapid sand filtration	
Slow sand filtration	Slow sand filtration	
Safety chlorination	Safety chlorination	

[a]Only applied when the water temperature exceeds 8–10°C.

which recently have shown up in The Netherlands will be discussed in relation to treatment.

OXIDATION TREATMENT IN WATER SUPPLY

Chlorine

Application of a chlorine treatment for disinfection purposes is an effective way for inactivating bacteria of hygienic significance present in water.[6,7] Rook[8] and Bellar[9] showed however that halogenated hydrocarbons are introduced in (drinking) water as a result of this treatment. Later on, a survey of the EPA in 1975 showed that approximately 50% of the volatile nonpolar compounds in drinking water are halogenated.[10] The question arose whether these compounds pose a health risk. The trihalomethanes (THM) in particular received major attention since one of these compounds (viz., chloroform) showed carcinogenic effects in animals.[11–13] Furthermore, it was shown in several drinking water studies that a chlorine treatment also is able to introduce and increase mutagenic activity[14–16] as well as nonvolatile organohalides.[17–18] An example of introduction of halogenated hydrocarbons is shown in Table 2.

Ozone

Ozone has been used for many decades in drinking water supply in Europe. It is a powerful disinfectant for bacteria, fungi, and viruses. A large quantity of organic matter, however, will impede the inactivation ability of ozone.[6,7] Ozone improves taste and odor of the water and oxidizes organics to more polar compounds. Studying the possible toxic effect of wastewater,

Table 2. Effect of a Chlorine Treatment (1 mg/L Cl$_2$) on the Level of Organohalides in Drinking Water

Source of Drinking Water	Organic Parameter[a]					
	Before Treatment			After Treatment		
	THM (μg/L)	VOX (nmol hal/L)[b]	AOX (μmol hal/L)[b]	THM (μg/L)	VOX (nmol hal/L)[b]	AOX (μmol hal/L)[b]
Groundwater	ND[c]	ND	0.6	28	450	2.5
Bank infiltrated water	ND	ND	1.0	5.2	18	2.8
Surface water	ND	6	1.1	70	1340	3.0

[a]THM = trihalomethanes; VOX = volatile organohalides; AOX = nonvolatile carbon adsorbed organohalides (sample is purged with N$_2$).
[b]Microcoulometric titration.
[c]ND = not detectable.

Gruener[19] showed that ozone was able to produce mutagenic effects. Kool et al.[20] summarized studies concerning mutagenic effects of ozone in drinking water treatment and concluded that in some cases ozone decreased, increased, or had hardly any influence on mutagenic activity. It was concluded that the varying results in the different ozone experiments very likely are due to differences in the composition of the raw water, process conditions, and methods of concentration, since large differences in this respect occur. It was also shown that ozone did not influence the level of volatile organohalides and decreased the level of nonvolatile carbon adsorbable organohalides (AOX).[20,21] Whether this is due to a conversion of adsorbable to nonadsorbable halogenated compounds is not yet clear. An example of an ozone treatment on the level of organohalides is shown in Table 3.

Chlorine Dioxide

Chlorine dioxide is effective for disinfection against bacteria and viruses. Compared to chlorine (weight base) it is more efficient at higher pH.[7] Therefore it is under investigation to see whether this agent can be used as an alternative disinfectant to chlorine, since several studies have shown[22–27] that chlorine dioxide hardly elevates the level of volatile organohalides. On the other hand, it was found[24,25] that, at a reduced pH, formation of AOX took place and that ClO$_2$ will be transformed to ClO$_2^-$,[22] which may pose a health risk.

High doses of chlorine dioxide, 5 to 15 mg/L ClO$_2$, are able to increase mutagenic activity.[28] When the level of chlorine dioxide is reduced (1 mg/L ClO$_2$), Kool et al.[21,29] showed that, depending on the type of water, a slight increase, a decrease, or no effect at all on mutagenic activity was observed. In contrast with chlorine the mutagenic activity (both a dose of 1 mg/L) is

Table 3.   Effect of an Ozone Treatment (3 mg/L O₃) on the Level of Organohalides in Drinking Water

| | Organic Parameter[a] | | | |
| | Before Treatment | | After Treatment | |
Source of Drinking Water	THM (μg/L)	AOX (μmol · hal/L)[b]	THM (μg/L)	AOX (μmol · hal/L)[b]
Surface water	ND	0.7	ND	0.2
Bank infiltrated river water	ND	0.2	ND	ND
Groundwater	ND	0.2	ND	ND

[a]THM = trihalomethanes; AOX = nonvolatile carbon adsorbed organohalides (sample is purged with N_2).
[b]Microcoulometric titration.

relatively small in comparison to the activity obtained after a chlorine treatment. Chlorine dioxide had hardly any effect on the volatile organohalides but increased the level of nonvolatile organohalides (AOX) in most types of water.[21,24,25,29] An example is shown in Table 4.

Considering the discussed oxidation treatment (chlorine, chlorine dioxide, ozone), the conclusion is justified that mutagenic activity and volatile and nonvolatile organohalides in general are introduced or are increased in substantial amounts by a chlorine treatment. For chlorine dioxide, mutagenic activity may be introduced or increased. The latter also counts for volatile and nonvolatile organohalides. When this happens, however, the mutagenic activity and the level of organohalides are substantially less than obtained with chlorine. Ozone also may introduce or increase mutagenic activity but seems to reduce the level of nonvolatile organohalides. One should bear in mind, however, that mutagenic activity can be removed effectively by granular activated carbon (GAC) filtration.[30-34] Volatile and nonvolatile organohalides can also be removed effectively by GAC filtration, but only for a relatively short time. By far the best results are obtained by GAC filtration preceded by an ozone treatment.[35,36]

CARCINOGENIC EFFECTS DUE TO ORGANICS IN DRINKING WATER

There is a concern over possible adverse health effects due to the presence of organic micropollutants and by-products in drinking water in particular when a chlorine treatment is applied. Many epidemiological studies in this respect have been carried out primarily in the United States.[37] Two recent case comparison epidemiological studies[38,39] have provided an indication of a risk associated with chlorinated water for colon and bladder cancer. Recently many organic data concerning drinking water have been reconsidered

Table 4. Effect of a Chlorine Dioxide Treatment (1 mg/L ClO$_2$) on the Level of Organohalides in Drinking Water

| Source of Drinking Water | Organic Parameter[a] | | | | | |
| | Before Treatment | | | After Treatment | | |
	THM (μg/L)	VOX (nmol hal/L)[b]	AOX (μmol hal/L)[b]	THM (μg/L)	VOX (nmol hal/L)[b]	AOX (μmol hal/L)[b]
Surface water	ND	ND	0.7	ND	7	1.4
Bank-infiltrated water	ND	ND	0.2	0.7	5	0.6
Groundwater	ND	ND	<0.1	ND	ND	0.8

[a]THM = trihalomethanes; VOX = volatile organohalides; AOX = nonvolatile carbon adsorbed organohalides (sample is purged with N$_2$).
[b]Microcoulometric titration.

in relation to cancer mortality, and it was found that there was, statistically, a highly significant relationship between gastrointestinal tract cancers and exposure to (chlorinated) organics in drinking water.[40]

Animal studies regarding the carcinogenicity of drinking water organics have been carried out in which mice (skin) and rats were exposed to organic concentrates prepared from drinking water. In the United States, chloroform-extracted activated carbon extracts showed negative[41] and positive results.[42] Dunham et al.[43] used the same procedure in two areas with a high and low incidence of bladder cancer. This study showed no induction of carcinogenic effects. In France, Hemon et al.[44] applied a chloroform extraction procedure of Cabridenc and Sdika[45] and showed cancer-promoting activity in a mouse skin test. Truhaut et al.[46] showed with drinking water of Paris (chloroform extraction) an increased tumor induction in rats and mice. Only a dose-response effect was observed in rats. Two mutagenic drinking water concentrates out of five obtained by reverse osmosis (RO) and XAD resins increased the number of papillomas in the presence of phorbol myristate acetate, proving initiating (carcinogenic) properties.[47] All samples (5 XAD and 5 RO concentrates) failed to be complete carcinogens, but the duration of the experiment (38 weeks) was rather short. Bull et al.[48,49] investigated whether in a mouse skin test carcinogenic effect occurred by disinfecting water with chlorine, chlorine dioxide, ozone, and chloramine. The results clearly showed that the composition and certain precursors present is the treated water greatly influenced the skin tumor incidence. Long-term carcinogenicity studies carried out in The Netherlands with rats showed that a mixture of 11 chlorinated hydrocarbons frequently detected in drinking water did not induce carcinogenic effects.[50] In another study whereby mutagenic organic concentrates prepared from chlorinated drinking water were tested, no carcinogenic effects were shown.[51] A summary of carcinogenic studies with complex organic mixtures from drinking water is shown in Table 5.

Table 5.   Summary of Carcinogenicity Studies with Organic Concentrates/Mixtures Prepared from Drinking Water

Year	Investigator(s)	Assay	Method of Concentration	Results
1954	Hueper et al.			Negative
1963	Hueper et al.	Mouse (skin)	Chloroform-activated carbon extracts	Positive (control?)
1967	Dunham et al.			Negative
1978	Hemon et al.	Mouse (skin)	Chloroform extraction	Positive (cancer promoting activity)
1979	Truhaut et al.	Mouse/rat	Chloroform extraction	Positive (control?)
1980	Bull et al.	Mouse	RO/XAD	Negative (exposure too short—38 weeks)
1985	Kool et al.	Rat	XAD	Negative
1985	Wester et al.	Rat	—[a]	Negative

[a]Mixture of 11 chlorinated hydrocarbons.

Many carcinogenic studies have been carried out or are underway with respect to chemicals found in drinking water, in particular to chemicals that are by-products of chlorination. Most of these data recently have been reviewed by Bull.[52] It is concluded that the data available at present are not suited for projecting the risk to humans from the use of materials in the treatment and distribution of drinking water. Identification of carcinogenic properties of by-products of disinfection other than the trihalomethanes, acrylamide, and coal tar paints, however, does indicate that the use of such material in contact with potable water requires closer scrutiny.

Considering the possible adverse health effects of organic contaminants in drinking water, it is reasonable to assume, as stated by Clark et al.,[40] that if carcinogenic effects are due to organics in drinking water, it results primarily from chlorination by-products rather than from exposure to synthetic organics. The recent study of Clark et al.[40] showed a highly significant relationship between gastrointestinal and urinary tract cancers and organics in drinking water. Since this kind of epidemiological study is not able to establish a causal relationship, long-term animal studies have to be carried out to prove carcinogenicity. The (long-term) carcinogenicity studies carried out so far do not, however, give firm support to the positive results obtained in the latest epidemiological studies.[38–40]

The conclusion therefore is that proving whether carcinogenic effects are due to the consumption of drinking water containing organic micropollutants is almost impossible unless many long-term animal studies are carried out.

Another approach to solve this dilemma is to accept the philosophy of disinfection of waterworks in The Netherlands:

> Postchlorinations should be omitted when the biological quality (hygienic aspects and aftergrowth) is sufficient and when the distribution system does not need protection.

This philosophy has been applied always for groundwater and for more than a century for the Waterworks in The Hague, which uses surface water as raw water source. In 1983, the Waterworks of Amsterdam reduced its safety chlorination[53] and finally stopped this treatment. Some chemical and biological parameters before and after stopping safety chlorination are shown in Tables 6 and 7.

Table 6. Level of Dissolved Organic Carbon (DOC) and Organohalides in Drinking Water Before (B) and After (A) Stopping Safety Chlorination

Raw Water Source	DOC (mg C/L)		AOC[a] (μg/L)		THM[b] (μg/L)		AOX[c] (μg Cl/L)	
	B	A	B	A	B	A	B	A
River-Dune water	1.9	1.9	10	< 6	12–18	ND[d]	50–65	10–12
Lake water	4.2	4.3	20	<12	20–22	ND	30–35	5–10

Source: Schellart.[53]
[a]AOC = assimilable organic carbon (microgram acetate equivalents/L).
[b]THM = trihalomethanes.
[c]AOX = nonvolatile carbon adsorbed organohalides.
[d]ND = not detectable.

Table 7. Level of Some Microbial Parameters in Drinking Water Before and After Stopping Safety Chlorination

	Microbial Parameters							
	Before				After			
Drinking Water Source	Coliforms[a]	Fecal Streptococci[a]	Plate Count (cfu/mL)		Coliforms[a]	Fecal Streptococci[a]	Plate Count (cfu/mL)	
			22°C	37°C			22°C	37°C
Pumping Station	ND[b]	ND	<11	<3	ND	ND	<11	<3
Tap	ND	ND	<11	<3	ND	ND	<11	<3

Source: Schellart.[53]
[a]500 mL tested.

From this experience, Schellart[53] concludes:

- From a hygienic point of view, chemical disinfection is not needed before drinking water is distributed.
- One should decrease the level of AOC (assimilable organic carbon),[54] since this category is responsible for regrowth of bacteria.
- Physical/chemical and mechanical treatment are preferred over oxidative treatments like chlorine and ozone. The latter hardly removes the organic carbon. On the contrary, it increases the AOC level.
- Slow sand filtration is an attractive alternative to final disinfection before distributing drinking water.

RECENT PROBLEMS IN DRINKING WATER

Aeromonas Bacteria

In May 1984, the Waterworks of The Hague discovered many atypical colonies on Teepol agar after membrane filtration. These colonies appeared, after isolating and typing, to be *Aeromonas hydrophila*. These *Aeromonas* spp. are accepted to be opportunistic pathogens in immunologically compromised hosts[55] but may also be enteric pathogens in immunologically normal patients.[56] The discovery of a few hundred *Aeromonas* bacteria in 500 mL drinking water led to an intensive investigation with respect to source and behavior of these bacteria during water treatment. For hygienic purposes, a safety chlorination treatment of 1 mg/L chlorine was introduced for about a month and finally reduced to 0.5 mg/L till the end of the year and then stopped. The investigation showed that *A. hydrophila*, *A. sobria*, and *A. caviae* were found throughout all stages of treatment. They were removed by slow sand filtration and, to a lesser extent, by dune filtration. Only one slow sand filter showed aftergrowth of these bacteria. After the filter was put out of order the situation clearly improved. In combination with a safety chlorination step during a limited period, the number of *Aeromonas* bacteria was greatly reduced. From this experience, a survey was started in The Netherlands to see whether *Aeromonas* bacteria were also present in other types of drinking water. Therefore the National Institute of Public Health and Environmental Hygiene sampled nearly all pumping stations in The Netherlands for *Aeromonas* bacteria.[57] The results of this survey are shown in Table 8.

The results clearly show that the majority of the drinking water samples did not contain *Aeromonas* bacteria. The results are questionable, however, since recently it was found that the presence of copper (60 µg/L) reduced completely the number of *Aeromonas* bacteria.[58] Since in pumping stations and distribution systems copper pipelines are often used, the results found so far may underestimate the actual number of *Aeromonas* bacteria in drink-

Table 8. *Aeromonas* Bacteria Detected in Drinking Water in The Netherlands

Number of Samples from Pumping Station	Number of Distribution Samples	*Aeromonas* Bacteria (cfu/100 mL)
136 (71%)	140 (71%)	ND[a]
40 (21%)	40 (20%)	1–20
9 (5%)	10 (5%)	21–50
8 (4%)	10 (5%)	>50

Source: Versteegh, During, Havelaar, and Koot.[57]
[a]ND = not detectable.

ing water due to exposure of copper. Therefore the sampling procedure has been modified by adding 25–50 mg Na_2EDTA per 500 mL.[58] Despite this phenomenon, relatively high numbers of *Aeromonas* have been found in drinking water prepared from groundwater in a few pumping stations (Table 9).[59]

Further epidemiological and ecological studies are underway to find out whether the presence of *Aeromonas* bacteria in drinking water poses a real health risk.

Table 9. *Aeromonas* Bacteria Detected in Drinking Water Prepared from Groundwater in The Netherlands (1986)

Number of Samples from Pumping Station	*Aeromonas* Bacteria[a] (cfu/100mL)
12	ND[b]
6	1–20
3	>200

[a]Mainly *A. hydrophila.*
[b]ND = not detectable.

Nitrate in Groundwater

The presence of nitrate in groundwater has received renewed attention because the level of nitrate dramatically is increasing. Subsequently the nitrate level in drinking water is increasing, too (Figure 1).

The increase of the level of nitrate in groundwater and drinking water is mainly due to high application rates of manure in sandy areas with intensive husbandry. The nitrogen in the manure is transformed into nitrate by microbiological processes. An example of this phenomenon in porewater from a water catchment area is shown in Table 10.

It is obvious that a large amount of nitrate is moving toward the groundwater level and will greatly deteriorate the groundwater quality in the near future. The actual situation with respect to the nitrate problem in vulnerable

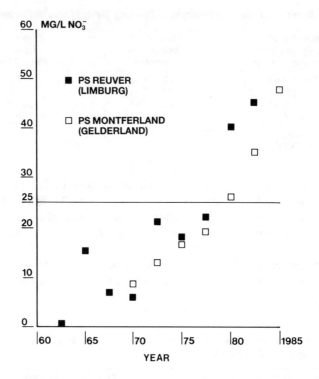

Figure 1.   Course of nitrate in drinking water (source groundwater) in areas with corn.

areas has been established.[60,61] A view of the course of nitrate and sulfate in drinking water of 34 pumping stations in the eastern part of The Netherlands is shown in Table 11.

From this table, it appears that in nine pumping stations (27%) nitrate is increasing and in four out of these nine pumping stations (12%) sulfate is also. In 11 pumping stations (32%), only sulfate is increasing. These results indicate that microbiological processes in soil limit the increase of nitrate by denitrification in which sulfur acts as substrate, and therefore the sulfate level will increase. This phenomenon has been described previously.[62] In the next few years, it is hoped the nitrogen load in water catchment areas will decrease to such an extent that nitrate pollution of groundwater will not further increase. In the meantime, the waterworks are obliged to reduce the nitrate level in drinking water, because it is very unlikely that in the next twenty years the nitrate level in groundwater will decrease significantly. It takes a long period before all nitrate has reached the saturated zone. A pilot plant study therefore has been started in which nitrate is removed by denitrification. Two denitrification processes are under investigation and have been

Table 10.   Level of Nitrate in Porewater in a Water Catchment Area with a Corn Culture

Sample Depth Below Surface (m)	Level of nitrate[a] (mg/L)
0–1	244
2–3	311
4–5	493
6–7	835
8–9	2278
10–11	367
12–13	521
14–15	495
16–17	274
18–19	256
20–21	293

[a]Nitrate level calculated on base on the presence of ammonia and nitrate (Bruyn[59]).

Table 11.   Change of Nitrate and Sulfate Level in Drinking Water Derived from Groundwater in Eastern Part of The Netherlands (1962–1984)

Number of Pumping Stations[a]	Course of Nitrate and Sulfate Level in Drinking Water
14	No increase of nitrate and sulfate
4	Increase of nitrate and sulfate Nitrate range: 4–18 mg/L Sulfate range: 22–79 mg/L
5	Increase of only nitrate Nitrate range: 8–48 mg/L Sulfate range: 10–25 mg/L
11	Increase of only sulfate Nitrate range: 0–20 mg/L Sulfate range: 16–90 mg/L

Source: Kool.[61]
[a]Production of 34 pumping stations is about 100×10^6 m^3/year.

described previously.[63–65] Some preliminary results of a sulfur/limestone denitrification process are shown in Table 12.

These results show that nitrate can be removed completely. On the other hand an increase in sulfate and hardness will occur, which limits the application of this denitrification treatment. Only drinking water with relatively low hardness and sulfate levels can be treated in this way. Denitrification with sulfur/limestone filtration seems a reliable process, since it has already been applied for more than a year without failures. Furthermore it appears that this process can be stopped easily and restarted. Drinking water which does not satisfy the hardness and sulfate conditions should use another denitrification process in combination with an ion exchange step. This process is also under investigation in the pilot plant study.[64]

Table 12. Nitrate Removal in Groundwater by Denitrification with Sulfur/Limestone Filtration

Chemical Parameters	Before Treatment (mg/L)	After Treatment (mg/L)
NO_3^-	65–70	0–25
O_2	6	0
HCO_3^-	60–65	140–150
SO_4^{2-}	20–30	110–140
pH	6.8–7.0	7.5

Source: Schippers, Kruithof, Mulder, and van Lieshout.[65]

SUMMARY AND CONCLUSIONS

Chlorine dioxide and ozone are preferred over chlorine with respect to formation of mutagenic activity and organohalides in drinking water treatment. In contrast with chlorine, both treatments occasionally introduce or increase mutagenic activity depending on the type of water. For ozone the process conditions are also an important factor. Formation of organohalides certainly takes place with chlorine, occurs in some cases with chlorine dioxide, but has not been seen with ozone. The production of organohalides by chlorine dioxide is substantially less than for chlorine. It seems that ozone decreases the level of organohalides (nonvolatiles).

Evidence of carcinogenic effects caused by chlorinated drinking water is rather weak. Although several epidemiological studies have shown an association between chlorinated drinking water and cancer mortality of the gastrointestinal and urinary tracts, no firm conclusions can be drawn in this respect. Only one long-term animal carcinogenicity study with drinking water supported this association. Questions, however, can be raised with respect to the control group in this positive study.

From mouse (skin) studies it appeared that organic concentrates prepared from chlorinated drinking water showed tumor-initiating properties. In order to obtain convincing evidence whether carcinogenic effects are caused by chlorinated drinking water, one approach is to do more long-term animal studies and specific epidemiological studies. Another approach is to discuss whether a safety disinfection is a necessary treatment in drinking water supply. In The Netherlands the philosophy becomes more and more accepted that *no chemical disinfection* has to be applied when *sufficient* physical, chemical, and biological barriers are present. This philosophy, which has been accepted by most waterworks in The Netherlands, has (recently) also been brought in practice by The Municipal Waterworks of Amsterdam (raw water source river Rhine) by stopping its safety chlorination. No hygienic problems occurred up till now.

Recently, "new" problems have shown up in drinking water supply in The

Netherlands. First, relatively high numbers of *Aeromonas* bacteria (an enteric pathogen) have been found in drinking water prepared from surface water and groundwater. Whether the presence of these numbers of bacteria pose a real health risk is under investigation. Second, due to high application rates of manure in areas with intensive husbandry, nitrate levels in ground- and drinking water are increasing drastically. Furthermore the level of nitrate in groundwater will rise in the next decade to such an extent that it is difficult, and in some cases almost impossible, to meet the drinking water standard of 50 mg/L nitrate without treatment. Therefore nitrate removal by two denitrification processes is now under investigation in a pilot plant study. A sulfur/limestone filtration process seems a very reliable denitrification step to remove nitrate, but is limited to drinking water with relatively low hardness and sulfate concentrations. Drinking water that does not satisfy both conditions should use a denitrification treatment in combination with an ion exchange step. Therefore application of denitrification processes in drinking water supply seems to be necessary in the near future.

REFERENCES

1. *Waterleiding Statistiek 1983, deel 1* (VEWIN, 1985, Rijswijk, The Netherlands).
2. *DeSamenstelling van het Rijnwater in 1984 en 1985* (RIWA, Postbus 8169, Amsterdam).
3. *Jaarverslag 1984* (DWL, 's Gravenhage).
4. Havelaar, A. H., and E. H. W. van Erne. "Escherichia Coli en Salmonella in de Grote Rivieren," RIV report 43/80. National Institute of Public Health and Environmental Hygiene, Bilthoven, The Netherlands (1979).
5. Van Olphen, M., and B. van de Baan. "Virologisch Onderzoek van Drinkwater en Oppervlaktewater in Nederland in de Periode 1978–1980," H_2O 15:626–631 (1982).
6. Kool, H. J. *Biological Indicators of Water Quality* (Chichester: John Wiley & Sons, Ltd., 1979), Chapter 17.
7. *Drinking Water Microbiology.* NATO/CCMS Drinking Water Pilot Project Series CCMS 128. EPA Office of Drinking Water, Washington, DC, (1984).
8. Rook, J. J. "Formation of Haloforms During Chlorination of Natural Waters," *J. Water Treatm. Exam.* 23:234–243 (1974).
9. Bellar, T. A., J. J. Lichtenberg, and R. O. Kroner, "The Occurrence of Organohalides in Chlorinated Drinking Water," *J. Am. Water Works Assoc.* 66:703–706 (1974).
10. "National Organic Reconnaissance Survey. Analysis of Tapwater from Five U.S. Cities for Volatile Organic Compounds," U.S. EPA staff report HERL, Cincinnati, OH (1975).
11. "Carcinogenesis Bioassay of Chloroform." NTIS no. PB 264018/AS (1976).
12. Bull, R. J. "Carcinogenic and Mutagenic Properties of Chemicals in Drinking Water," *Sci. Total Env.* 47:385–413 (1985).

13. *Chloroform*. IARC Monographs on the Evaluation of Carcinogenic Risk of Chemicals to Man, vol. 20. IARC, Lyon (1979), 401–427.

14. Cheh, A. M., J. Skochdopole, P. Koski, and L. Cole. "Non-Volatile Mutagens in Drinking Water: Production by Chlorination and Destruction by Sulphite," *Science* 207:90-92 (1980).

15. Loper, J. C. "Mutagenic Effect of Organic Compounds in Drinking Water," *Mutation Res.* 76:241–268 (1980).

16. Kool, H. J., F. van Kreijl, and B. C. J. Zoeteman. "Toxicity Assessment of Organic Compounds in Drinking Water," *CRC Crit. Rev. Environ. Control* 12:307–357 (1982).

17. Oliver, B. G. "Chlorinated Non-Volatile Organics Produced by the Reaction of Chlorine with Humic Materials," *Can. J. Res.* 11:21–22 (1978).

18. Glaze, W. M., F. Y. Salek, and W. Kintsley. "Characterization of Non-Volatile Halogenated Compounds Formed During Water Chlorination," in *Water Chlorination: Environmental Impact and Health Effects, vol. 3*, R. L. Jolley, W. A. Brungs, and R. B. Cumming, Eds. (Ann Arbor, MI: Ann Arbor Science Publishers, Inc., 1980), 99.

19. Gruener, N. "Mutagenicity of Ozonated Recycled Water," *Bull. Environ. Contam. Toxicol.* 20:522–526 (1978).

20. Kool, H. J., J. Hrubec, C. F. van Keyl, and G. J. Piet. "Evaluation of Different Treatment Processes with Respect to Mutagenic Activity in Drinking Water," *Sci. Total Environ.* 47:229–256 (1985).

21. Kool, H. J., and J. Hrubec. "The Influence of an Ozone, Chlorine and Chlorine Dioxide Treatment on Mutagenic Activity in Drinking Water," *Ozone Sci. Engin.* 8:217–234 (1986).

22. Miltner, R. J. V. "The Effect of Chlorine Dioxide on Trihalomethanes in Drinking Water," M.S. Thesis, University of Cincinnati, Cincinnati, OH (1976).

23. Fiessinger, F., Y. Richard, A. Montiel, and R. Musquere. "Advantages and Disadvantages of Chemical Oxidation and Disinfection by Ozone and Chlorine Dioxide," *Sci. Total Environ.* 18:245–261 (1981).

24. Kuhn, W., and H. Sontheimer, "Treatment: Improvement or Deterioration of Water Quality?" *Sci. Total Environ.* 18:219–233 (1981).

25. Stevens, A. A. "Reaction Products of Chlorine Dioxide," *Environ. Health Pers.* 46:101–110 (1982).

26. Chou, B. M., and P. Roberts. "Halogenated By-Products, Formation by ClO_2 and Cl_2," *J. Environ. Eng. Div., ASCE* 107:109–118 (1981).

27. Bull, R. J. "Toxicological Problems Associated with Alternative Methods of Disinfection," *J. Am. Water Works Assoc.* 74:642–648 (1982).

28. De Greef, E., J. C. Morris, C. F. van Kreyl, and C. F. H. Morra. "Health Effects in the Chemical Oxidation of Polluted Water," in *Water Chlorination: Environmental Impact and Health Effects, vol. 3*, R. L. Jolley, W. A. Brungs, and R. B. Cumming, Eds. (Ann Arbor, MI: Ann Arbor Science Publishers, Inc., 1980), 913.

29. Kool, H. J., C. F. van Kreijl, and J. Hrubec. "Mutagenic and Carcinogenic Properties of Drinking Water," in *Water Chlorination: Chemistry, Environmental Impact and Health Effects, vol. 5*, R. L. Jolley, R. J. Bull, W. P. Davis, S. Katz, M. H. Roberts, Jr., and V. A. Jacobs, Eds. (Chelsea, MI: Lewis Publishers, Inc., 1985), 187.

30. Pendygraft, G. N., F. E. Schlegel, and M. J. Huster. "Organics in Drinking Water: Maximum Contaminant Levels as an Alternative to the GAC Treatment Requirement," *J. Am. Water Works Assoc.* 71:174–183 (1979).
31. Kool, H. J., C. F. van Kreijl, E. de Greef, and H. J. van Kranen. "Presence, Introduction, and Removal of Mutagenic Activity in Drinking Water in The Netherlands," *Environ. Health Pers.* 46:207–214 (1982).
32. Van der Gaag, M. A., A. Noordsij, and J. P. Oranje. *Progress in Chemical and Biological Research, vol. 109,* (New York: Alan R. Liss, Inc., 1982).
33. Hrubec, J., C. F. van Kreijl, C. F. M. Morra, and W. Slooff. "Treatment of Municipal Waste Water by Osmosis and Activated Carbon-Removal of Organic Micro-Pollutants and Reduction of Toxicity," *Sci. Total Environ.* 27:71–88 (1983).
34. Monarca, S., J. R. Meier, and R. J. Bull. "Removal of Mutagens from Drinking Water by Granular Activated Carbon," *Water Res.* 17:1015–1026 (1983).
35. Kruithof, J. C., A. Noordsij, L. M. Puijker, and M. A. van der Gaag. "Influence of Water Treatment Processes on Formation of Organic Halogens and Mutagenic Activity by Postchlorination," in *Water Chlorination: Chemistry, Environmental Impact and Health Effects, vol. 5,* R. L. Jolley, R. J. Bull, W. P. Davis, S. Katz, M. H. Roberts, Jr., and V. A. Jacobs, Eds. (Chelsea, MI: Lewis Publishers, Inc. 1984), 1137.
36. Van der Gaag, M., J. C. Kruithof, and L. M. Puijker. "The Influence of Water Treatment Processes on the Presence of Organic Surrogates and Mutagenic Compounds in Water," *Sci. Total Environ.* 47:137–153 (1985).
37. *Epidemiological Studies of Cancer Frequency and Certain Organic Constituents of Drinking Water. Review of Recent Literature Published and Unpublished,* National Academy of Sciences, Washington, DC (1979).
38. Cragle, D. L., C. M. Shy, R. J. Struba, and E. J. Siff. "A Case-Control Study of Colon Cancer and Water Chlorination in North Carolina," in *Water Chlorination: Chemistry, Environmental Impact and Health Effects, vol. 5,* R. L. Jolley, R. J. Bull, W. P. Davis, S. Katz, M. H. Roberts, Jr., and V. A. Jacobs, Eds. (Chelsea, MI: Lewis Publishers, Inc., 1979), 153.
39. Cantor, K. P., R. Hoover, P. Hartge, T. J. Mason, D. T. Silverman, and L. I. Levin. "Drinking Water Source and Risk of Bladder Cancer: A Case-Control Study," *Water Chlorination: Chemistry, Environmental Impact and Health Effects, vol. 5,* R. L. Jolley, R. J. Bull, W. P. Davis, S. Katz, M. H. Roberts, Jr., and V. A. Jacobs, Eds. (Chelsea, MI: Lewis Publishers, Inc., 1985), 143.
40. Clark, R. M., J. A. Goodrich, and R. A. Deininger. "Drinking Water and Cancer Mortality," *Sci. Total Environ.* 53:153–172 (1986).
41. Hueper, W. C., and C. C. Ruchhoft. "Carcinogenic Studies on Adsorbates of Industrially Polluted Raw and Finished Water Supplies," *Arch. Ind. Hyg.* 9:488–495 (1954).
42. Hueper, W. C., and W. W. Payne. "Carcinogenic Effects of Adsorbates of Raw and Finished Water Supplies," *Am. J. Clin. Path.* 39:475–481 (1963).
43. Dunham, L. J., R. W. O'Gara, and R. B. Taylor. "Studies on Pollutants from Processed Water: Collection from Three Stations and Biologic Testing for Toxicity and Carcinogenesis," *Am. J. Public Health* 57:2178–2185 (1967).
44. Hemon, D., P. Lazar, R. Cabridenc, A. Sdika, B. Festy, C. Gerin-Roze, and

I. Chouroulinkov. "Micropollution organique des eaux destinées a la consommation humaine," *Rev. Epidem. Sante Publ.* 26:441–450 (1978).

45. Cabridenc, R., and A. Sdika. "Quelques aspects de l'extraction et de l'identification des micropollutants des eaux," *Tech. Sci. Munic.* 70:285–288 (1975).

46. Truhaut, R., J. C. Gak, and Cl. Graillot. "Recherches sur les risques pouvant resulter de la pollution chimique des eaux d'alimentation-I," *Water Res.* 13:689–697 (1979).

47. Robinson, M., J. W. Glass, D. Cmehill, R. J. Bull, and J. Orthoefer. *Short Term Bioassays in the Analysis of Complex Environmental Mixtures II* (New York: Plenum Press, 1980).

48. Bull, R. J. "Health Effects of Alternative Disinfectants and Their Reaction Products," *J. Am. Water Assoc.* 72:299–303 (1980).

49. Bull, R. J., M. Robinson, J. R. Meier, and J. Stober. "Use of Biological Assay Systems to Assess the Relatively Carcinogenic Hazards of Disinfection By-Products," *Environ. Health Pers.* 46:215–227 (1982).

50. Wester, P., C. A. van der Heyden, A. Bisschop, G. J. van Esch, R. C. C. Wegman, and Th. de Vries. "Carcinogenicity Study in Rats with a Mixture of Eleven Halogenated Hydrocarbons Drinking Water Contaminants," *Sci. Total Environ.* 47:427–432 (1985).

51. Kool, H. J., F. Kuper, H. van Heuningen, and J. H. Koeman. "A Carcinogenicity Study with Mutagenic Organic Concentrate of Drinking Water in The Netherlands," *Fd. Chem. Toxic.* 23: 79–85 (1985).

52. Bull, R. J. "Carcinogenic and Mutagenic Properties of Chemicals in Drinking Water," *Sci. Total Environ.* 47:385-413 (1985).

53. Schellart, J. "Disinfection and Bacterial Regrowth: Some Experiences of the Amsterdam Waterworks Before and After Stopping the Safety Chlorination," *Water Supply* 4:217–225 (1986).

54. Van der Kooy, D., A. Visser and W. A. M. Hynen. "Determining the Concentration of Easily Assimilable Organic Carbon in Drinking Water," *J. Am. Water Works Assoc.* 74:540–545 (1982).

55. Davis, W. A., J. G. Kane, and V. P. Garagusi. "Human Aeromonas Infection: A Review of the Literature and a Case Report of Endocarditis," *Medicine* 57:267–277 (1978).

56. Gracey, M., V. Burke, and J. Robinson. "Aeromonas Associated Gastro Enteritis," *Lancet* 11:1304–1306 (1982).

57. Versteegh, J. F. M., M. During, A. H. Havelaar, W. Koot. "Aeromonas spp. in het Nederlands Drinkwater: Een Orienterend Onderzoek," RIVM rapport nr. 84111002, Bilthoven, The Netherlands (1986).

58. Van Oorschot, R. "Onderzoek naar het Toxisch Effect van Koper op Aeromonas spp. en naar de Mogelijkheden van de Defoxificatie door Toevoeging van EDTA," Rapport Duinwaterleiding—'s Gravenhage, Hoofd Afdeling Laboratorium Augustus (1986).

59. Bruyn, J. "Drinkwaterkwaliteit en Bemesting: Nitraat Problemen in Oost Gelderland," H_2O 7: 502–503 (1984).

60. Van Beek, C. G. E. M., D. van de Kooy, P. C. Noordam, and J. C. Schippers. "Nitraat en Drinkwatervoorziening," KIWA mededeling 84 KIWA NV, Rijswijk, The Netherlands (1984).

61. Kool, H. J. *De Wet Bodembescherming Ver. Milieurecht— Sectie Milieuchemie* (Tjeenk Willink, Zwolle, The Netherlands, 1986), 45.
62. Kolle, W., P. Werner, O. Strebel, and J. Bottcher. "Denitrification by Pyrite in a Reducing Aquifer," *Vom Wasser* 61:125–147 (1983).
63. Blecon, G. "Denitrification autotrophique par Thiobacillus denitrificons sur soufre-aspects microbiologique et mise au point technologique," These Rennes (1985).
64. Hoek, J. P., and A. van den Klapwijk. "Nitraat Verwijdering uit Grondwater," *H₂O* 18:57–62 (1985).
65. Schippers, J. C., J. C. Kruithof, F. G. Mulder, J. W. van Lieshout, "Nitraat Verwijdering met Langzame Zwavel/Kalksteen Filtratie," *H₂O* 20:30–31 (1987).

CHAPTER 2

Understanding the Response to Environmental Risk Information

Joanne Vining, Institute for Environmental Studies, University of Illinois at Urbana-Champaign, Urbana, Illinois

INTRODUCTION

As the list of toxic and hazardous substances in drinking water has lengthened, the problem of communicating the risks attributable to those substances to the public has become critical. Even when information about risks is available it is often inconclusive,[1] and may involve controversy due to conflicting scientific opinions.[2] Indeed, even the apparently simple task of monitoring the quality of water can be quite difficult due to the small concentrations of some toxic and hazardous substances and due to their transient nature. Identifying the human health consequences and risks is even more problematic because causal statements are difficult to achieve in human health research, which relies on inferences from correlational epidemiological studies or on extrapolations from clinical trials with nonhuman animals. Increasingly, suspected risks must be communicated under conditions of uncertainty before more conclusive evidence is available.

When the public is presented with information about the risks of environmental hazards, the variety of responses is often bewildering. People are generally very concerned about the risks they face. At the same time, they can be remarkably sensitive to some risks and insensitive to others. An individual who accepts the risks of smoking may nonetheless become quite agitated when informed that trichloroethylene has been found in his or her drinking water. Or, flood victims who proclaim the horrors of a flood, may in the next breath announce their intentions to rebuild their homes in the flood plain.

Despite these apparent inconsistencies, there are consistencies in people's responses to risk. In general, people consistently use a variety of strategies to simplify gathering and processing information to make a decision. Because these strategies are semiautomatic they are occasionally used inappropriately, and much has been made of results which are outwardly irrational or illogical. However, this sort of mental autopilot is effective in the vast majority of situations, and it is important to recognize the functional value of these strategies. Life is complex and risky, and our predecessors who spent a lot of time agonizing over myriad details of daily judgments and decisions probably did not live long enough to become our ancestors. Thus, we take mental and emotional shortcuts and rely on intuition to tell us when we are safe and when we must take precautions.

In the remainder of this chapter I will summarize some of these strategies, beginning with the application of basic psychological principles to the problem of risk perception. In the second section I will present some of the more robust (i.e., replicable) empirical findings about the way people think about risk, and third, discuss a few of the newer findings about emotional and intuitive factors in risk perception that psychologists and other social scientists have recently identified. The final section will present some recommendations for communicating risk information.

TWO BASIC PSYCHOLOGICAL PRINCIPLES APPLIED TO RISK PERCEPTION

The process of perceiving and evaluating risk is orderly and susceptible to the same laws as other forms of judgment and behavior. Two basic psychological principles, derived from the learning theory of Skinner and psychoanalytic theory of Freud, provide some insight into the way people may respond to risk information.

Behaviorism: Short-Term Gain vs Long-Term Pain

The central principle of behaviorism (also operant conditioning or reinforcement theory) is that the consequences of behavior determine its frequency in the future. Any behavior which is rewarded will increase in frequency, while unrewarded or punished behaviors will presumably decrease in frequency. A primary corollary to this principle is that the short-term consequences of behavior are much more important than long-term consequences in determining our behavior. People constantly conduct informal cost-benefit analyses, weighing the value of one behavior or action over another. When the object of a change in behavior is to benefit in the long term while suffering in the short, people will almost invariably choose to avoid short-term suffering. So, for example, when individuals or communities are confronted with the opportunity of treating water at great expense to avoid cancer in 30 years, or paying for remodeling immediately to reduce radon exposure, which may reduce the life span by two years, many may choose to accept the long-term risk so that the funds will be available for present needs or wants.

This principle also has important implications for communicating risk information. Often risks are expressed as reductions in life expectancy. Thus, the tradeoff is often expressed in terms of happiness now versus misfortune at the end of life. Given that people focus primarily on immediate rather than remote eventual consequences, it is no wonder that these statistics are often unheeded.

Defending Against Conflict and Stress

Evaluating risky information is complex and involves difficult trade-offs that people are reluctant to make. For example, deciding whether to expose one's family to radon gas or to allocate money for preventive remodeling may cause an individual to question the value he or she places on family health versus some other use of the money. This kind of evaluation is uncomfortable and stressful for most people, and their initial response is often anxiety or distress. One way to reduce the stress of these emotional states is to obtain the appropriate information and begin preventive or corrective action. However, when an individual lacks the emotional, mental, or financial resources to cope with the decision process, a defense mechanism may be used to shield against the anxiety or distress. Although defense mechanisms usually involve self-deception and reality distortion, they can be adaptive means of allaying anxiety when combined with other coping mechanisms. However, when defense mechanisms are used often and to the exclusion of other coping strategies, the solutions they provide may be less than optimal.

There are several varieties of defense mechanisms. Denial is the simplest of defense mechanisms. It involves screening out stressful reality by simply ignoring or refusing to acknowledge what one sees, thinks, or feels. Thus, an individual might ignore a potential health threat completely, or deny its importance.[3]

A second defense mechanism, which is often combined with denial, is rationalization. Through rationalization an individual avoids unpleasant thoughts or emotions by creating an alternative reality in which his or her behavior or thought is justified, usually through faulty logic. For example, an individual might decide that everything causes cancer, so it doesn't matter what she eats or drinks. Or, someone might deny the importance of risk information by announcing that "you can prove anything you want with statistics." Rationalization also may ease an individual's disappointment in his or her inability to solve a problem or reach a goal. Thus, lack of preparation for a disastrous event such as a flood may be rationalized by deciding that it was a freak occurrence that no one could have foreseen.

THOUGHTFUL STRATEGIES

Evaluating risk information and arriving at a decision as to the proper course of action is often a complex and arduous process. It often involves thinking about unpleasant eventualities such as disease or death. Furthermore, the information upon which one must act is often inconclusive and sometimes conflicting. An individual must assess the quality of the information and its source, and make judgments about the level of risk he or she faces and about the value of various alternative actions. Because these thoughts, judgments, and actions can be aversive and stressful, people often use a variety of cognitive or intellectual strategies to ease the burden. In general, these strategies involve avoiding cognitive effort by simplifying the information, the problem and its solution, and the alternatives for action.

Reducing Complexity

There is a limit to how much information people can deal with at once. Miller[4] proposed that individuals can manage only seven plus or minus two discrete items at once. The volume of information associated with most risk judgment problems is invariably greater than this, and usually has the added dimensions of uncertainty and complexity. When incoming information becomes too complex, conflicting, or voluminous, the individual often uses one of several strategies to reduce the complexity of the evaluation and decisionmaking process.[5]

One way to reduce the volume and complexity of incoming information is to sort items into categories. This leaves only the summary classifications to manage. For example, people will sometimes try to put risk information in absolute terms by asking whether a substance is either safe or not safe. By thus eliminating the gray areas of possibility or probability, people simplify the input upon which they must then act.[3,5]

Another way to simplify the input to a problem is to eliminate ambiguity in causal relationships. For example, an acquaintance's cancer could be due to any number of causes, one of which might be a carcinogen that is suspected to be present in local drinking water. The ambiguity and complexity of this situation may be eased by drawing the causal link between the hazard and the disease, and presuming that the cancer was caused by contamination of water. With this assumption taken as fact, an individual could then act to protect him or herself to prevent further contamination or to remove the contaminant from the water supply. Of course, depending on the person's opinions regarding the source or nature of the pollution, the causal relationship could be posed differently. It is not uncommon, for example, for factory workers to discount the hazardous nature of chemicals or substances manufactured by companies upon which their jobs depend. In this case, the causal agent might be presumed to be genetic factors or other sources of contamination.

These are two examples of means for simplifying the complexity of information upon which judgments and decisions must be made. Nisbett and Ross[6] or Kahneman et al.[7] may be consulted for a more comprehensive discussion of these and other strategies.

The "Don't Confuse Me with the Facts" Phenomenon

A second way of avoiding cognitive effort is to come to closure on a problem, make a decision, and abide by it even in the face of conflicting opinions or evidence. As Fischoff[5] noted, once peoples' minds are made up, it can be very difficult to change them. Coming to closure on a problem by making a decision allows people to forget much of the complexity of the information from which the decision was derived, and to remember only the decision. This saves cognitive effort by relieving the individual of the task of looking for contrary information. Most people guard against self-doubt and thus do not expect to find contrary information once they have reached a conclusion.

Perhaps more important, strongly held attitudes influence how further information is stored and interpreted.[8,9] Typically, new information that conforms with initial views is assessed as valuable, while information that contests our beliefs is considered inaccurate, unconvincing, or biased. Thus, if an individual believes that a local corporation is causing pollution and accompanying disease in the community, it is likely that information to the contrary will be largely discredited or ignored and confirming information heeded.

The problems that surround these phenomena are embodied in a recent problem with groundwater contamination in Tucson, Arizona. Allegedly, sometime in the 1950's, the Air Force and a local aircraft manufacturing company disposed of trichloroethylene (TCE) in containers that have since leaked. The TCE has reportedly found its way into the groundwater, and a number of wells are contaminated. Before the Air Force marshalled its considerable public information efforts, a local newspaper conducted an informal epidemiological study and reported the results. According to a series of news accounts, the number of cancers in areas with contaminated wells was higher than areas with uncontaminated wells. Although these reports were beneficial in getting the state to conduct formal epidemiological studies, unfortunately many people made up their minds about the pollution on the basis of the news accounts.

The exact effect of being the first to provide information and concern is unknown. However, in this example, in spite of the outcome of scientific epidemiological studies, the Air Force is in the unenviable position of correcting first impressions. If scientific studies support the link between TCE and disease, the Air Force will need to correct the impression that they were not concerned or interested enough to provide information first. Since the newspaper broke the story first, any action the Air Force takes will look like an afterthought or a response to public outcry and pressure. If a link between the pollution and disease is not found, the Air Force will have to try to change peoples' minds about the nature of the problem. In either case, the people involved have constructed beliefs and attitudes toward the problem on the basis of the first available information, and those attitudes will probably be very difficult to change.

Experience vs Facts and Figures

Even when a human health risk may be accurately and clearly specified (e.g., with a known death or tumor rate and clearly identifiable causal agent), people sometimes make errors in estimating the probability that those risks will occur. Lichtenstein et al.[10] proposed that the availability heuristic (originally identified by Tversky and Kahneman[11]) explains some regularities in these errors. The availability heuristic is a judgment strategy by which people use information that readily comes to mind. If information or memories regarding a particular hazard are readily available to the individual's judgmental process, his or her estimate of the frequency of that hazard will be higher than an estimate for hazards with less readily available information.

Studies of risk perception have found a number of reasons why information, memories, or events become available, or salient. Slovic et al.[12] found that estimates of death rates for infrequent hazards are often higher than the actual rates. Conversely, if a death rate is high, people often underestimate its frequency. Thus, for example, people overestimate their risk for death from botulism or terrorism, and underestimate the risk for heart disease and lung cancer. Slovic et al. proposed that some of the risks were overestimated because they were associated with dramatic or sensational events, such as natural disasters, which are easily brought to mind.

Another important factor which influences what readily comes to mind is personal experience. Personal experience is a major source of readily available material that is limited only by memory. Thus, people might feel that their drinking water is safe because they have been drinking it all their lives. Or, even if all of their direct experiences with air travel are safe, people may feel safer in their cars because they have a much larger number of safe experiences in their cars. The situation is much more complex than this, of course. The risk of dying in a plane crash is associated with feelings of lack of control, dread, and vivid television footage of flaming wrecks (the availability of emotional material is discussed below). However, because our personal experiences are highly memorable, and therefore available, we are much more likely to rely on what we know from past experience than what we know from actuarial tables in evaluating the risks we face.

Scientific Notation, Large and Small Numbers, Statistics and Probability

An additional problem with most numerical presentations of risk information is that understanding scientific notation, large and small numbers, statistics, and probability involves a sizable cognitive effort for most people. The
judgment and decisionmaking literature is replete with examples of miscon-
ceptions, errors, and biases in perceiving probability accurately (e.g., Nisbett and Ross[6]). For example, people often misinterpret the meaning of the
100-year flood, a statistical representation of the likelihood that a flood of a
certain magnitude will occur. Thus, flood plain residents will describe the
horror of the flood they just experienced, and then announce that it's a good
thing that it will be another ninety-nine years before it happens again. It is
easy to think that this example would only be characteristic of rather naive
individuals. However, even statistically literate scientists who should know
better often fall prey to errors of inference. For example, Tversky and
Kahneman[13] showed that their statistically sophisticated subjects placed more
faith than was warranted in the extent to which small samples represented the
populations from which they were drawn.

Using statistics, probability, and scientific notation to present data thus
becomes a problem when people are trying to simplify their lives. However,
if people have some difficulty with specifying the exact odds that they will
die from a relatively unlikely event, they nonetheless manage to conceive of
a rough idea of the risks they face. Although Slovic et al.[12] showed that
people overestimate infrequent causes and underestimate frequent causes of
death, it is important to note that the rank order of the probabilities was
perceived accurately even though actual estimates were imprecise. Despite
the fact that most people cannot specify the exact probabilities of death from
cancer or botulism, almost everyone understands very well that the former
is much more of a hazard than the latter and are quite capable of making finer
distinctions as well.[14] Increasingly, psychologists are focusing on the ade-
quacy and function of thought and behavior rather than on prescriptions for
precision and accuracy. As March[15] (p. 593) noted, "If behavior that apparently deviates from standard procedures of calculated rationality can be
shown to be intelligent, then it can plausibly be argued that models of
calculated rationality are deficient not only as descriptors of human behavior
but also as guides to intelligent choice."

These issues indicate that the way risk information is presented is critical.
People tend to be very sensitive to the way a risky choice is framed.[16] For
example, consider the outcome of the following pair of alternatives for presenting the risk for lowering a standard for water quality.

If a standard for water quality is lowered:
a. your chance of contracting cancer will change from 1 in 1,000,000 to 1 in 500,000.
b. your chance of contracting cancer will double.

It is not difficult to see that an individual viewing the second of these alternative formulations would be more likely to become alarmed and to act to prevent the standard change. As it is obvious that the way information is presented can manipulate perceptions of the problem, this comparison begs the question of ethics in presenting risk information. Fischoff[5] has suggested that information should be presented in more than one form so that people have a more complete understanding of the risk in question.

A second issue in presenting risk information is the fact that people rarely concern themselves with alternative or amplifying information that is not presented. For example, if a risk is expressed in terms of death rate, people will probably not think in terms of toxic reactions other than death, reductions in life span, or the probability of birth defects, carcinogenicity, or mutagenicity (Wilson and Crouch[17]). Looking for what is not there adds to the complexity of the problem. When only one outcome is used to communicate risk information, any omissions will probably not be perceived. Similarly, it is unusual for people to question the source of the data that are presented. For example, the Human Exposure dose/Rodent Potency dose (HERP) scale developed by Ames et al.[18] summarizes the relative risks of various natural and manufactured substances. When presented with this scale, which implies that eating peanut butter contaminated with aflatoxin is more risky than drinking water contaminated with chloroform, most people would not ask about the uncertainty involved in risk estimates, which amount to inferences from studies of animals, or about interaction effects among chemicals.

INTUITIVE AND/OR FEELING STRATEGIES

The way people respond to risk information is influenced by a number of factors, alone and in combination. Although studies of cognitive or intellectual responses to risk information have predominated in the literature for the past ten to fifteen years, the emotional and intuitive properties of the response to risk are increasingly being examined. These properties, although important, have not been studied as often because of difficulties in verbalizing and measuring emotion and intuition.

Risk Is an Emotional Issue

In general, the way people feel affects the way they think. Emotion or mood affects the way information is retrieved from memory, the way new information is perceived and processed, and the way decisions are made. Bower[19] has provided evidence that a specific mood state may influence the ease with which past events associated with similar moods are retrieved from memory. People who feel sad tend to remember sad things and attribute sadness things to other objects, events, and people, and people who are feeling hostile may attribute hostile feelings to others. Thus, a hostile audience at a public hearing for a serious environmental problem may interpret the attitudes, motives, and behavior of the officials conducting the meeting as hostile, whether they actually are hostile or not. Understanding this will not change the hostility, but it may make officials better prepared for the response of the public.

As discussed above, the extent to which memories are available influences their use in present decisionmaking. An intense emotional response to events may increase the availability of those events for subsequent decisions. For example, Fischoff et al.[20] factor analyzed ratings of riskiness of a variety of events and found that emotional factors such as dread, lack of control, and the catastrophic potential of events influence judgments of riskiness.

Johnson and Tversky[21] studied the extent to which specific fears may influence general estimates of risk. In a series of experiments they found that people who read news accounts of specific hazardous events gave increased estimates of risk for a variety of risky events. For example, after reading a story about homicide, subjects increased their estimates of the frequency of deaths due to homicide and for all other risks that they were asked to evaluate. Johnson and Tversky concluded that the mood that is induced by a news report has an extensive influence on risk estimates for related and unrelated events. Thus, someone who has responded fearfully to recent information about a plane crash may become more anxious in response to subsequent information about contaminants in water than someone who began the evaluation of the water contamination problem in a cheerful or neutral frame of mind.

Emotions also affect more specific judgments. For example, Clore[22] used hypnosis to induce either fear or anger in subjects. After they were brought out of the hypnotic state, subjects provided risk estimates (the frequency of death due to a variety of risks), and made blame attributions (the extent to which a person in a hypothetical situation was to blame for a problem). The emotions that subjects experienced had very specific effects on risk estimates and blame attributions: fear influenced subjects' risk estimates but not blame attributions; anger influenced blame attributions but not risk estimates. Positive feelings can also influence the extent to which people are willing to accept risk. However, the effects of positive mood seem to interact with the nature of the risk under evaluation and with the nature of the task. Isen and Patrick[23] found that people in a positive mood were more willing than controls in a neutral mood to take a chance on a low-risk bet, but not on a high-risk bet. However, when evaluating a hypothetical situation, elated subjects were more likely to accept a riskier situation.

These studies indicate that emotion can have specific and generalized effects on various judgmental tasks, including estimates of risk. Although further studies need to be done, the information available indicates that emotion creates a context for memory, and judgments, and decisions. Emotion is thus a factor to be reckoned with in interpreting the public response to risk information.

Appearances Are Important

The appearance or attractiveness of an object or event may influence how safe or risky it seems. Anyone who has tried to convince someone that water which has benign, but visible, dissolved particles is safer than water which is clear but hazardous knows that perceptions of safety and risk are correlated with perceptions of aesthetics or attractiveness. Schroeder and Anderson[24] examined this relationship in studies of the attractiveness and safety of urban parks and recreation areas. Their data and more recent preliminary data (Vining et al.[25]) indicate that many people find that safe places are attractive and vice versa. The direction of this mutual influence is not clear, however: it is as yet unknown whether perceived riskiness or safety influences attractiveness, or whether attractiveness influences perceptions of risk or safety. Furthermore, these studies have examined the relationship between safety and attractiveness of places, and not of substances, events, or objects. Although it is probable that these findings will transfer to other situations, it is important to determine that issue empirically.

These studies also point to the importance of individual differences in the perception of risk. Anderson and Schroeder[24] noted that peoples' past experiences with outdoor areas influence the relationship between their perceptions of safety and attractiveness. People from predominantly rural backgrounds rated wooded natural areas higher for both safety and attractiveness, while urban residents preferred more open, developed parks. These findings may also generalize to problems of the safety and attractiveness of clean water and air.

Understanding the response to environmental risk information will involve considering the aesthetics of situations, events, and objects as well as variations in individual responses to the presentation of information. Ruckelshaus (cited in Weinberg[26]) has noted that there has been a shift in emphasis on visible and demonstrable pollution (smog, dead lakes) to potential and/or invisible or imperceptible problems. This shift will add to the complexity of communicating risks and understanding the response of the public.

CONCLUSION AND RECOMMENDATIONS

This chapter has presented some of the generalizations which can be made about people's responses to environmental risk information. Understanding the response to risk information is complex and difficult because of the number and complexity of the variables involved. To begin with, information about environmental hazards and risks is generally imprecise and inferential in nature. Information often seems to have a character of its own, embodied by the source of the message, and the medium and format of transmission. Secondly, each individual brings his or her own intellectual or emotional strategies, personal characteristics, memories, and life background to the judgment and decision process. Finally, the situation or context in which judgments and decisions are made influences the process and the outcome. For example, the emotional, social, or cultural context may influence the perception of risk and the selection of alternative actions.

The psychological and sociological literature on risk is large and growing rapidly, yet there is still a lot to know. Communicating risk is a notoriously complex and difficult task,[2,27] and there are few definitive generalizations that are warranted from the studies which have been done to date. However, some recommendations can be made concerning the communication of environmental risk information to the public.

First, it is important to allow the complexity of risk information to enter into the judgment process. Fischoff[5] has noted that, despite the fact that people will try to avoid cognitive effort, it is best not to try to protect them from complex problems. Evaluating risky information and making decisions about alternative actions is too important to economize on input. Moreover, the controversy and uncertainty that is invariably involved with risk assessments should not be disguised or hidden.[28] It is best to keep complexity alive and present new angles or ways of thinking to jolt the judgment process and prevent premature categorization of information or assumptions. As a simple example, in order for people to understand information about changes in death rates due to changes in a water standard, it would be advisable to present the absolute rate (e.g., a change from one to two in five thousand) as well as the change per se (e.g., double or triple the existing rate), in addition to the level of uncertainty about the change.

Second, while preserving complexity, information should be provided in a format that people can understand and manage. Endless lists of probabilities expressed in scientific notation are precise, but usually not comprehensible to the average observer. Providing digestible information entails using examples which are relevant to the target audience's background and using a scale that is comprehensible or related to the audience's experience. For example, death rates could be put in terms of an individual's local population. It might be easier to understand the chances of contracting giardiasis from the local water supply in terms of the number of people in the local community who might be expected to acquire the disease rather than from a national average, or from a probability figure. One should always resist the temptation to simplify too much, however. As has been noted, the HERP scale[18] permits the comparison of the riskiness of various substances without providing an indication of the complexity of the information, the questionable basis for extrapolating from animal to human responses, and the problem of responses to interactions among chemicals. This scale is thus a good example of digestible but misleading information.

Third, it is probably best to be the first to make a balanced and comprehensible presentation of the risks people are asked to accept. In particular, if the problem is likely to be inflamed or sensationalized by others, it is important to express concern and interest before the story is reported elsewhere.

Finally, it is important to recognize the influence of emotion and intuitive factors and individual differences in preferences and judgmental styles. To the extent that is practical, an understanding of these factors should be incorporated into any communication of environmental risk information. For example, when conducting a public hearing where the audience is likely to be feeling hostile, it is important to realize that similar hostility may be attributed to those conducting the hearing. The point is not to try to manipulate or defuse feelings, but to understand how they may influence the perceptions and decisions of the recipients as well as the providers of information about environmental risks.

REFERENCES

1. Thomas, L. M. "Risk Communication: Why We Must Talk About Risk," *Environment* 28(2):4–5,40 (1986).
2. Hammond, K. R., B. F. Anderson, J. Sutherland, and B. Marvin. "Improving Scientists' Judgments of Risk," *Risk Anal.* 4(1):69–78 (1984).
3. Slovic, P., B. Fischoff, and S. Lichtenstein. "Rating the Risks," *Environment* 21(3):14–20,36–40 (1979).
4. Miller, G. A. "The Magical Number Seven Plus or Minus Two: Some Limits on Our Capacity for Processing Information," *Psych. Rev.* 63:81–97 (1956).
5. Fischoff, B. "Managing Risk Perceptions," *Issues Sci. Technol.* 3:83–96 (1985).
6. Nisbett, R., and I. Ross. *Human Inference: Stategies and Shortcomings of Social Judgment* (Englewood Cliffs, NJ: Prentice-Hall, Inc., 1980).
7. Kahneman, D., P. Slovic, and A. Tversky, Eds. *Judgment Under Uncertainty: Heuristics and Biases* (Cambridge: Cambridge University Press, 1982).
8. Slovic, P. "Perception of Risk," *Science* 236:280–285 (1987).
9. Tversky, A., and D. Kahneman. "Belief in the Law of Small Numbers," *Psych. Bull.* 76:105–110 (1971).
10. Lichtenstein, S., P. Slovic, B. Fischoff, M. Layman, and B. Combs, "Judged Frequency of Lethal Events," *J. Exp. Psych: Human Learning and Memory* 4:551–578 (1978).

11. Tversky, A., and D. Kahneman. "Availability: A Heuristic for Judging Frequency and Probability," *Cognitive Psych.* 5:207–232 (1973).

12. Slovic, P., B. Fischoff, and S. Lichtenstein. "Facts versus Fears: Understanding Perceived Risk," in *Judgment Under Uncertainty: Heuristics and Biases,* D. Kahneman, P. Slovic, and A. Tversky, Eds. (Cambridge: Cambridge University Press, 1982).

13. Tversky, A., and D. Kahneman. "Belief in the Law of Small Numbers," *Psych. Bull.* 76:105–110 (1971).

14. Fischoff, B., and D. MacGregor. "Judged Lethality: How Much People Seem to Know Depends Upon How They Are Asked," *Risk Anal.* 3(4):229–236 (1983).

15. March, J. G. "Bounded Rationality, Ambiguity, and the Engineering of Choice," *Bell J. of Econ.* 9:587–608 (1978).

16. Slovic, P. "Informing and Educating the Public About Risk," *Risk Anal.* 6(4):403–415 (1986).

17. Wilson, R., and E. A. C. Crouch. "Risk Assessment and Comparisons: An Introduction," *Science* 236:267–270 (1987).

18. Ames, B. N., R. Magaw, and L. S. Gold. "Ranking Possible Carcinogenic Hazards," *Science* 236:271–279 (1987).

19. Bower, G. H. "Mood and Memory," *Am. Psych.* 36:129–148 (1981).

20. Fischoff, B., P. Slovic, S. Lichtenstein, S. Read, and B. Combs. "How Safe is Safe Enough: A Psychometric Study of Attitudes Towards Technological Risks and Benefits," *Policy Sciences* 9:127–152 (1978).

21. Johnson, E. J., and A. Tversky. "Affect, Generalization, and the Perception of Risk," *J. Pers. and Soc. Psych.* 45:20–31 (1983).

22. Clore, G. L. "The Cognitive Consequences of Emotion and Feeling," paper presented at the meeting of the American Psychological Association, Los Angeles, CA, August, 1985.

23. Isen, A. M., and R. Patrick. "The Effect of Positive Feelings on Risk Taking: When the Chips are Down," *Org. Beh. and Human Perf.* 31:194–202 (1983).

24. Schroeder, H. W., and L. M. Anderson. "Perception of Personal Safety in Urban Recreation Sites," *J. Leis. Res.* 16(2):178–194 (1984).

25. Vining, J., A. Ebreo, and M. Walker. "Individual Differences in Perceived Safety and Aesthetics in Urban Parks and Forests." Unpublished working paper (1987).

26. Weinberg, A. M. "Science and Its Limits: The Regulator's Dilemma," *Issues Sci. Technol.* 3:59–72 (1985).

27. Sharlin, H. I. "EDB: A Case Study in Communicating Risk," *Risk Anal.* 6(1): 61–68 (1986).

28. Whittemore, A. S. "Facts and Values in Risk Analysis for Environmental Toxicants," *Risk Anal.* 3(1):23–33 (1983).

SECTION II

Volatile Organic Compounds in
Treated Drinking Water

CHAPTER 3

Abiotic Transformation of Halogenated Organic Compounds
II. Considerations During Water Treatment

William J. Cooper and Rose A. Slifker, Drinking Water Research Center, Florida International University, Miami, Florida

Jeffrey A. Joens, Chemistry Department, Florida International University, Miami, Florida

Ombarek A. El-Shazly, Chemistry Department, Suez Canal University, Ismailia, Egypt

INTRODUCTION

Reports of groundwater and surface water contamination are quite common. In many cases, the contaminants reported are halogenated organic compounds.[1-4] As a result, there has been increasing concern about contamination of drinking water when these source waters are used.[3-7] Another area of concern is the possibility of changes occurring in these compounds by chemical reactions during water treatment. The purpose of this study was to examine the abiological reactions of halogenated ethanes, accelerated by the high pH encountered in the lime softening process or in the water distribution system, where the water is often maintained at pH 9 to prevent corrosion. Under these conditions, reactions can continue at rates accelerated relative to that of the source water (pH generally around 7). In large distribution systems it is not uncommon to have water in the system (prior to use) for up to six days, allowing ample time for reactions to occur.

Two of the common abiological reactions of halogenated ethanes (in aqueous solution) are elimination

$$-\overset{\displaystyle |}{\underset{\displaystyle H}{C}} - \overset{\displaystyle |}{\underset{\displaystyle X}{C}} - \quad\longrightarrow\quad \overset{\diagdown}{\diagup}C = C\overset{\diagup}{\diagdown} \quad + \quad HX \qquad (1)$$

and substitution

$$R - X + H_2O \longrightarrow R - OH + HX \qquad (2)$$

The extent to which either reaction 1 or 2 occurs depends upon the halogenated ethane and the reaction conditions.

Several studies have been reported which discuss the aqueous reactions of various halogenated alkanes under environmentally relevant conditions.[8-17] The factors affecting the rate of reaction of halogenated alkanes in aqueous solution have been extensively discussed. In a critical review of hydrolysis reactions, Mabey and Mill[18] identified the experimental parameters affecting reaction rates as pH, temperature, ionic strength, buffer composition, and catalysis by metal ions or other substances. A complete study of the abiotic homogeneous reactions of a halogenated alkane requires characterization of the effect of each of the above factors on the rate and mechanism of reaction. In particular, reactions need to be studied over a wide range of pH to separate neutral reaction processes from acid or base mediated reactions, and over a range of temperatures sufficient to determine the Arrhenius parameters for the reaction rate constant and test for non-Arrhenius behavior. Comparisons of the rate of reaction in different buffer systems can be used to decide if a particular reaction is a specific acid/base promoted reaction, dependent on hydrogen and/or hydroxide ion concentration, or a general acid/base reaction.[19] The effect of ionic strength on the reaction rate can be used as a probe for the reaction mechanism,[20] while information on metal ion or other catalysis is important in applying laboratory results to conditions found in the environment.

The focus of the study reported herein, a part of a larger study, was to examine several compounds under conditions likely to be encountered during water treatment and distribution. The compounds studied in detail were 1,1,1,2-tetrachloroethane, 1,1,2,2-tetrachloroethane, and pentachloroethane. These compounds were chosen because they appear to react primarily by E2 elimination to form a single product, 1,1,2-trichloroethene (TCE) for the tetra-substituted ethanes, and tetrachloroethene (PCE) from pentachloroethane.

In addition to the above compounds, limited studies were conducted using 1,1,1-trichloroethane, one of the most common chemicals found in contaminated groundwaters.[21] This compound has been shown to undergo both elimination and substitution.[1,22-24] It appears that both of these reactions occur by neutral processes in a ratio of 1:3 independent of temperature and pH.[24]

We have recently described the elimination reaction of 1,1,2,2-tetrachloroethane through the pH range of 5 to 9.[25] The reaction leads to the formation of 1,1,2-trichloroethene, with 100% conversion, and was found to be a second-order homogeneous reaction, first order in base and in 1,1,2,2-tetrachloroethane. Additional studies are in progress to determine the effect of ionic strength on the rate of the elimination reaction.[26] The results obtained thus far are consistent with and extend the previous measurements of Mill and coworkers for the compound.[27,28]

Two other previous studies have investigated the reaction of 1,1,1,2-tetrachoroethane.[17,27] It appears that the major pathway is elimination yielding TCE. The elimination reaction involving pentachloroethane has also been shown to occur and to lead to tetrachloroethene as the major reaction product.[17] It is possible that substitution reactions on the C1 carbon (substituted with three Cl) could occur; however, this has not been reported to date.

Thus, it appears that these compounds may be affected during water treatment. The question is to what extent and under what conditions these reactions occur.

METHODS AND MATERIALS

The waters studied were either tap water from two locations in Miami, or water from two treatment plants, sampled after lime softening and prior to recarbonation. These samples ranged in pH from 9.0 to 10.5 and were lower in ionic strength than the buffered distilled waters used to study the reactions in the laboratory. After water samples were obtained, they were immediately returned to the laboratory and the experiments initiated. In one case, source water at pH 7 was also studied. Table 1 provides a summary of the pertinent water quality characteristics of the samples used in this study.

The analytical methods and kinetic analysis have been described previously.[25]

RESULTS AND DISCUSSION

Kinetics of Reactions

As expected, with the exception of 1,1,1-trichloroethane, the elimination reaction pathway accounted for the major reaction products in the compounds. Table 2 summarizes the data obtained for the four compounds at 30°C. Under conditions of constant pH, the loss of 1,1,1,2-tetrachloroethane, 1,1,2,2-tetrachloroethane, and pentachloroethane obeyed first-order kinetics. The reaction of pentachloroethane was studied only at pH 7 because even at 30°C, the reaction occurred more quickly at elevated pH

Table 1. Water Chemistry of Samples Used for Hydrolysis of Halogenated Organic Chemicals

Sample	Alkalinity (mg/L CaCO$_3$)	Hardness (mg/L CaCO$_3$)	pH	Conductivity (mmhos)	Total Organic Carbon (mg/L)
Tap water					
1	30.6	61	9.20	0.245	5.4
2	43.4	63	9.55	0.235	12.5
1A	32.0	52	9.10	NR[a]	NR
1B	28.2	54	9.20	0.227	4.4
2A	47.7	71	8.79	0.318	11.6
Lime softening plant					
1	44.5	64	10.27	0.290	5.5
2	55.7	74	9.87	0.345	11.3
1A	45.0	59	10.10	0.285	4.7
2A	45.0	59	9.98	0.298	11.7
Influent					
1	218	229	7.10	NR	13.5
2	229	226	7.20	NR	30.2

than could be studied by our experimental technique. The results in Table 2 are divided into two groups, 1,1,1-trichloroethane and 1,1,1,2-tetrachloroethane, where the half-life for reaction is on the order of days, and 1,1,2,2-tetrachloroethane and pentachloroethane, where the half-life is on the order of hours.

Table 3 summarizes data obtained at several values of temperature and pH for pentachloroethane. These data were obtained in 0.03 M phosphate buffer. Under all conditions the loss of pentachloroethane was followed by a corresponding increase in PCE (Figure 1). To evaluate the dependence of the elimination reaction rate constant on temperature, the Arrhenius equation was used.

$$k_e = A \exp(-E_a/RT) \qquad (3)$$

The results are shown in Figure 2. Based on these data the following Arrhenius parameters were obtained:

$$\log_{10}A = 14.27 \pm 0.33$$
$$E_a = -77.5 \pm 2.0 \text{ kJ mol}^{-1}$$

Our results for the elimination reaction of pentachloroethane are in fair agreement with those of Walraevens and coworkers,[17] who found $\log_{10}A = 12.50$ and $E_a = -63.6$ kJ mol^{-1}. Using their parameters results in a value of k_e at 25°C that is about a factor of 2 smaller than that found in the present study. The reason for the discrepancy is not known at present, but may be

Table 2. Summary of Experimental Data for Homogeneous Abiotic Hydrolysis of Several Halogenated Ethanes at 30°C

Sample	pH	Percent Degradation	k_{obsd} (da^{-1})	k_e (M^{-1} da^{-1})	$t_{1/2}$ (day)
		1,1,1-Trichloroethane			
Tap water 1	9.20	70	0.0055		126
Tap water 2	9.55	65	0.0049		141
Lime softening plant 1	10.27	67	0.0052	N/A[a]	133
Lime softening plant 2	9.87	67	0.0052		133
		1,1,1,2-Tetrachloroethane			
Tap water 1	9.20	29	0.00191	82.7	362
Tap water 2	9.55	34	0.00237	45.8	292
Lime softening plant 1	10.27	86	0.0132	48.5	53
Lime softening plant 2	9.87	72	0.00768	71.0	90
			(hr^{-1})	(M^{-1} sec^{-1})	(hr)
		1,1,2,2-Tetrachloroethane			
Tap water 1A	9.10	>90	0.227	4.32	3.0
Tap water 1B	9.20	>90	0.230	3.95	3.0
Tap water 2A	8.79	>90	0.108	3.33	6.4
Lime softening plant 1A	10.10	>90	2.11	3.34	0.33
Lime softening plant 2A	9.98	85	1.28	2.56	0.54
		1,1,1,2,2-Pentachloroethane			
Influent plant 1	7.10	83	0.175	264	4.0
Influent plant 2	7.20	85	0.189	227	3.7

[a]Both substitution and elimination are major pathways for this compound.

Table 3. Summary of Experimental Data for the Homogeneous Elimination Reaction of Pentachloroethane[a]

Temp. (°C)	pH (0.03)	No. of Points	k_{obsd} (hr^{-1})	k_e (M^{-1} s^{-1})
20	8.04	6	0.08525	31.3
30	6.76	6	0.0262	86.6
30	8.06	6	0.436	72.3
40	5.46	7	0.00742	248
40	6.76	6	0.120	201
40	8.08	6	2.20	177
50	5.45	6	0.0318	590
50	6.76	7	0.574	522
60	5.43	6	0.122	1360
60	6.76	7	2.25	1170
70	5.45	5	0.491	3150

[a]All data taken in 0.030 M phosphate buffer.

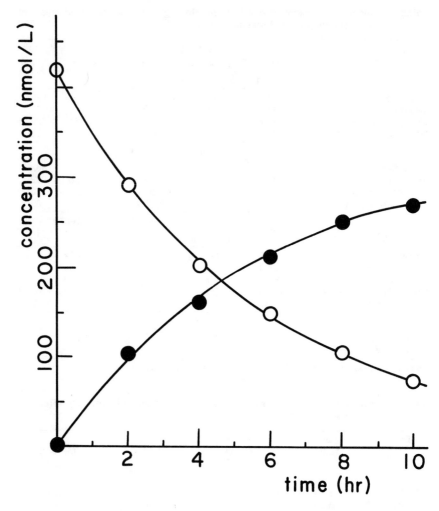

Figure 1. The disappearance of pentachloroethane (○) and appearance of tetrachloroethene (●) in 0.030 M phosphate buffer, pH = 7.2, T = 30°C, influent to plant A. The solid lines are based on calculations using the overall equation from this research.

due to differences in the experimental approaches. The major differences in the experimental methods are

1. we used much lower concentrations of halogenated ethane than in the Walraevens study;
2. we determined the concentration of pentachloroethane and the PCE elimination product directly by gas chromatography, while Walraevens and coworkers followed the reaction by measuring the concentration of chloride

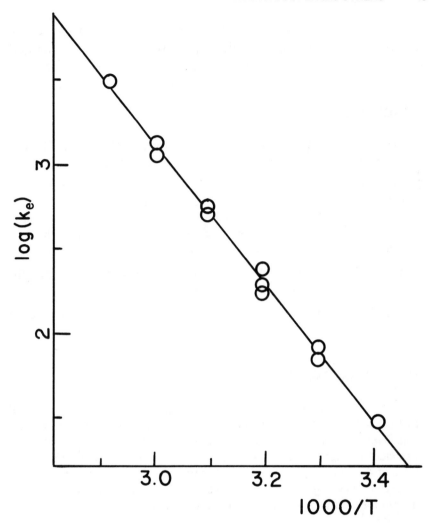

Figure 2. Arrhenius plot of the elimination of pentachloroethane, in 0.030 M phosphate buffers pH ~ 5.5 to 8.

ion produced by the elimination reaction, using conductivity measurements on the reaction solution; and

3. in general, the pH we used was lower than those used by Walraevens, reflecting more closely the pH expected to be encountered in the environment.

Effect of Metals

The elimination reaction of 1,1,2,2-tetrachloroethane was studied at pH 6.85 in 0.10 M phosphate buffer at 70°C, in the presence of individual heavy metals. The metals, each at 1 mg L^{-1}, were Fe(III), Al(III), Zn(II), Sb(II), Mn(II), Pb(II), Cd(II), Ni(II), Co(II), and Hg(II). In no case did the addition of these metals significantly affect the rate of the elimination reaction.

CONCLUSIONS

1. The normal practice of lime softening and water distribution at pH 9 will promote significant abiotic transformation of chlorinated ethanes.

2. Based on the present results and previous studies, elimination is the only important reaction pathway for 1,1,1,2-tetrachloroethane, 1,1,2,2-tetrachloroethane and pentachloroethane at room temperature and neutral or basic pH. The elimination reactions in all cases are first order in halogenated ethane and first order in hydroxide ion.

3. The disappearance of 1,1,1-trichloroethane is independent of pH over the limited range of pH examined in the present study.

4. Heavy metals in solution do not affect the rate of the elimination reaction of 1,1,2,2-tetrachloroethane at neutral pH.

ACKNOWLEDGMENTS

We would like to thank the U.S. Environmental Protection Agency, Grant R-811473–01–0, and the Drinking Water Research Center, Florida International University for support of this research. This research has not been subjected to Agency review and therefore does not necessarily reflect the views of the Agency, and no official endorsement should be inferred.

Special thanks go to Nahid Golkar, Mehrzad Mehran, and Mostafa Mehran for their technical assistance on the project.

REFERENCES

1. Pearson, C. R., and G. McConnell. "Chlorinated C1 and C2 Hydrocarbons in the Marine Environment," *Proc. Royal Soc. London* B189:305–332 (1975).
2. G. W. Page. "Comparison of Groundwater and Surface Water for Patterns and Levels of Contamination by Toxic Substances," *Environ. Sci. Technol.* 15:1475–1481 (1981).
3. National Research Council. *Groundwater Contamination* (Washington, DC: National Academy Press, 1984).
4. Vogel, T. M., C. S. Criddle, and P. L. McCarty. "Transformations of Halogenated Aliphatic Compounds," *Environ. Sci. Technol.* 21:722–736 (1987).
5. L. Fishbean. "Potential Halogenated Industrial Carcinogenic and Mutagenic

Chemicals. II. Halogenated and Saturated Hydrocarbons," *Sci. Total Environ.* 11:163–195 (1979).

6. J. C. Petura. "Trichloroethylene and Methyl Chloroform in Groundwater: A Problem Assessment," *J. Am. Water Works Assoc.* 73:200–205 (1981).

7. National Academy of Science. "Drinking Water and Health," (Washington, DC: National Academy Press), vol. 1 (1977); vol. 3 (1980); vol. 4 (1982); vol. 5 (1983).

8. Robertson, R. E., R. L. Heppolette, and J. M. W. Scott. "A Survey of Thermodynamic Parameters for Solvolysis in Water," *Can J. Chem.* 37:803–824 (1959).

9. Laughton, P. M., and R. E. Robertson. "Solvolysis in Hydrogen and Deuterium Oxide," *Can. J. Chem.* 37:1491–1497 (1959).

10. Leffek, K. T., R. E. Robertson, and S.Sugamori. "Heat Capacity of Activation in Hydrolytic Displacement from a Tertiary Carbon," *J. Am. Chem. Soc.* 87:2097–2101 (1965).

11. Moelwyn-Hughes, E. A., R. E. Robertson, and S. Sugamori. "The Hydrolysis of t-Butyl Chloride in Water: Temperature Dependence of the Energy of Activation," *J. Chem. Soc.* 1965:1965–1971 (1965).

12. Queen, A., and R. E. Robertson. "Heat Capacity of Activation for the Hydrolysis of 2,2-Dihalopropanes," *J. Am. Chem. Soc.* 88:1363–1365 (1966).

13. Heppolette, R. L., and R. E. Robertson. "Effect of α-Methylation on the Parameters Characterizing Hydrolysis in Water for a Series of Halides and Sulfonates," *Can. J. Chem.* 44:677–684 (1966).

14. Robertson, R. E. "Solvolysis in Water," *Prog. Phys. Org. Chem.* 4:213–280 (1967).

15. Blandamer, M. J., R. E. Robertson, E. Ralph, and J. M. W. Scott. "Kinetics of Solvolytic Reactions," *J. Chem. Soc. Faraday Trans. I* 79:1289–1302 (1983).

16. Schwarzenbach, R. P., W. Giger, C. Schaffner, and O. Warner. "Groundwater Contamination by Volatile Halogenated Alkanes: Abiotic Formation of Volatile Sulfur Compounds Under Anaerobic Conditions," *Environ. Sci. Technol.* 19:322–327 (1985).

17. Walraevens, R. P. Trouillet, and A. Devos. "Basic Elimination of HCl From Chlorinated Ethanes," *Int. J. Chem. Kinetics* 6: 777–786 (1974).

18. Mabey, W., and T. Mill. "Critical Review of Hydrolysis of Organic Compounds in Water Under Environmental Conditions," *J. Phys. Chem. Ref. Data* 7:383–415 (1978).

19. Perdue, E. M., and N. E. Wolfe. "Prediction of Buffer Catalysis in Field and Laboratory Studies of Pollutant Hydrolysis Reactions," *Environ. Sci. Technol.* 17:635–642 (1983).

20. Hammett, L. P. *Physical Organic Chemistry*, 2nd ed. (New York: McGraw-Hill Book Company, 1970), 187–219.

21. Westrick, J. J., J. W. Mello, and R. F. Thomas. "The Groundwater Supply Survey," *J. Am. Water Works Assoc.* 76:52–59 (1984).

22. Vogel, T. M., and P. L. McCarty. "Rate of Abiotic Formation of 1,1-Dichloroethylene from 1,1,1-Trichloroethane in Groundwater," *J. Contaminant Hydrol.* 1:299–308, 1987.

23. Cline, P. V., J. J. Delfino, and W. J. Cooper, "Hydrolysis of 1,1,1-Trichloroethane; Formation of 1,1-Dichloroethene," Proceedings of the NWWA/API

Conference on Petroleum Hydrocarbons and Organic Chemicals in Ground Water—Prevention, Detection and Restoration, Houston, TX, November 12–14, 1986.

24. Haag, W. R., and T. Mill. "Effect of Subsurface Sediments on Hydrolysis of Haloethanes and Epoxides," *Environ. Sci. Technol.* (submitted).

25. Cooper, W. J., M. Mehran, D. J. Riusech, and J. A. Joens. "Abiotic Transformations of Halogenated Organics. I. Elimination Reaction of 1,1,2,2-Tetrachloroethane and Formation of 1,1,2-Trichloroethene," *Environ. Sci. Technol.* 21:1112–1114 (1987).

26. Joens, J. A., W. J. Cooper, and R. A. Slifker. "Abiotic Transformations of Halogenated Organics. IV. Ionic Strength and Buffer Effects on the Elimination Reaction of 1,1,2,2-Tetrachloroethane," (in preparation).

27. Mabey, W. R., V. Barich, and T. Mill. "Hydrolysis of Polychlorinated Alkanes," Preprint Extended Abstracts, Division of Environmental Chemistry, American Chemical Society National Meeting, Washington, DC (1983), 359–361.

28. Haag, W. R., T. Mill, and A. Richardson. "Effect of Subsurface Sediment on Hydrolysis Reactions," Preprint Extended Abstracts, Division of Environmental Chemistry, American Chemical Society National Meeting, Anaheim, CA (1986), 248–253.

CHAPTER 4

Transformation Kinetics of 1,1,1-Trichloroethane to the Stable Product 1,1-Dichloroethene

Patricia V. Cline and Joseph J. Delfino, University of Florida, Gainesville, Florida

INTRODUCTION

1,1-Dichloroethene (1,1-DCE) was one of the five most frequently detected volatile organic compounds, other than trihalomethanes, in finished drinking water supplies according to a survey by the U.S. Environmental Protection Agency.[1] 1,1-Dichloroethene is a highly reactive, flammable liquid that is primarily used in the production of copolymers with vinyl chloride or acrylonitrile. Emissions occur during manufacturing, shipping, and production; however, these represent less than 1% of the produced 1,1-DCE.[2] The common occurrence of this compound as a groundwater contaminant cannot be entirely explained by its production and usage patterns.

One source of 1,1-DCE occurs during the abiotic degradation of 1,1,1-trichloroethane (TCA). The production of TCA is more than three times the production of 1,1-DCE, and unlike 1,1-DCE, it is an end use product, indicating that emission to the environment is essentially equivalent to the production.[2] The presence of 1,1-DCE is typically associated with the presence of other alkyl halides. Since 1,1-DCE is more toxic than TCA,[2] the conversion to 1,1-DCE in groundwater can increase the toxicity of the water supply.

The association of 1,1-DCE with TCA can be seen more dramatically in field data from sites that have high levels of solvents. A summary of volatile organic compounds (VOCs) in Arizona's groundwater[3] states that of the six most commonly detected VOCs, only three (1,1,1-trichloroethane, trichloroethene, and tetrachloroethene) are used in large quantities. The presence of 1,2-dichloroethene isomers and 1,1-dichloroethane, particularly with frequent detections of vinyl chloride, suggests anaerobic biodegradation.[4,5] Selected locations show very high levels of 1,1-DCE in association with TCA and frequently little evidence of biodegradation (Table 1). The primary source of 1,1-DCE at these locations appears to be the chemical degradation of TCA, prompting questions as to the rate of formation and stability of the 1,1-DCE.

Two products are formed during the abiotic degradation of TCA. The elimination product is 1,1-DCE, while the substitution or hydrolysis product is acetic acid (Figure 1). Factors which may influence the rate and pattern of the hydrolysis of alkyl halides include pH, temperature, and ionic strength.

Previous chlorinated solvent degradation studies have been reviewed. Dilling et al.[6] performed reactivity studies on selected compounds, including TCA. Estimated rates were based on four measurements for each of two sets of ampules; one set was maintained in the laboratory and a second set was kept outdoors in Michigan. The same estimated rate (half-life of approximately 6 months) was reported for each experiment. Products were not measured. The hydrolysis of TCA in seawater was reported by Pearson and McConnell.[7] A half-life of 39 weeks was estimated for TCA at 10°C with the predominant reaction being dehydrochlorination to 1,1-DCE. Walraevens et al.[8] examined the degradation of TCA in 0.5, 1.0, and 2.0 M sodium hydroxide solutions. Elimination was not observed, while sodium acetate was shown by infrared analysis to be the sole reaction product. 1,1-DCE was assumed to be stable under all experimental conditions.

Selected experiments performed in our laboratory were reported previously. They demonstrated the formation of 1,1-DCE at elevated temperatures in groundwater samples and phosphate buffers, with the elimination pathway accounting for 20–30% of the overall degradation.[9] Vogel and McCarty[10] monitored the degradation of TCA and formation of 1,1-DCE in water at pH 7 and 20°C. The TCA half-life at 20°C was estimated to range between 2.8 and 19 years. Haag and Mill[11] report approximately 22% conversion of TCA to the elimination product, with an extrapolated half-life of 350 days at 25°C. The studies reported in the literature provide some conflicting information on the rate and transformation processes of TCA, making extrapolation of rates to field sites or during treatment processes difficult. The meas-

Table 1. **Maximum Concentrations (μg/L) of VOCs Detected at Selected Sites in Arizona**

Site	TCA	1,1-DCE	TCE	1,2-DCE
1	630	3,320	13,000	20
2	490	1,320	9	—
3	9,800	10,400	410	933
4	98	206	139	106

Source: Graf.[3]

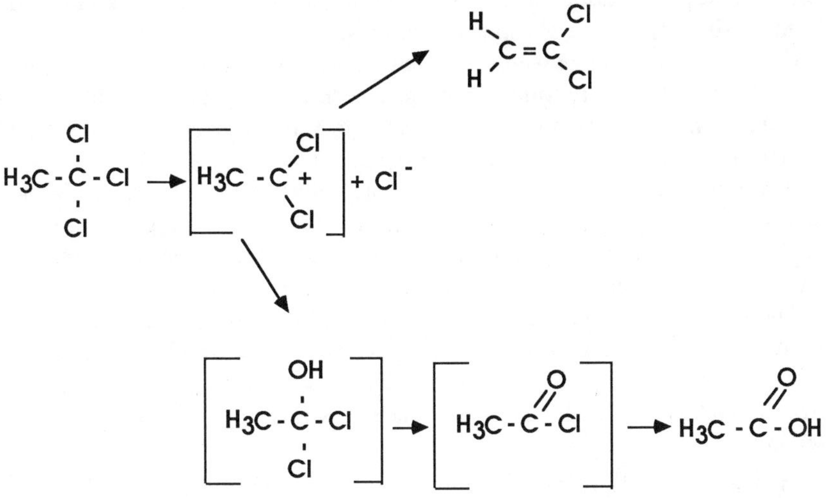

Figure 1. Abiotic degradation pathways for 1,1,1-TCA, with proposed S$_N$1 mechanism.

urement of the overall rate of degradation of TCA and the proportion of products formed, as well as the definition of parameters that influence these factors, were evaluated in our laboratory.

MATERIALS AND METHODS

Reagent-grade chemicals (Fisher Scientific) were used to prepare buffers and standard solutions. Phosphate solutions ($0.05\,M$) were prepared at pH 4.5, 7.0, and 8.5 by mixing stock solutions and monitoring the pH with a Fisher Accumet Model 230A pH meter. Solutions of $0.05\,M$ potassium dihydrogen phosphate and $0.05\,M$ potassium hydrogen phosphate were prepared using distilled deionized water. Equal molar volumes of these solutions were used for the pH 7.0 buffer. The phosphate solutions at pH 4.5 (potassium dihydrogen phosphate) and pH 8.5 (potassium hydrogen phosphate) required minor pH adjustment, using $0.05\,M$ phosphoric acid or potassium hydroxide solutions.

Stock standard solutions of TCA and 1,1-DCE were prepared in methanol at approximately 1 mg/mL concentrations. Working standards were prepared by spiking approximately 5 µL of the stock solution into 10 mL of distilled deionized water. Aliquots of 100–500 µL of the standard solution were used to prepare standard curves for the response of the gas chromatograph to the concentration of analyte.

Seawater samples were obtained from the coastal Atlantic Ocean near

Ormond Beach, Florida. Samples were 0.45 μm-filtered and subsequently handled similarly to the phosphate solutions.

Groundwater samples were collected from two monitoring wells located at a site in Orlando, Florida, which had been contaminated by chlorinated solvents. The water samples were purged to remove existing solvents and interfering substances, then filter-sterilized.

Approximately 6.6 mL of the phosphate solutions, seawater, or distilled deionized water were added to 5-mL (nominal volume) glass ampules (Wheaton Scientific), thus minimizing the head space. The ampules were plugged with cotton and autoclaved at 121°C for 15 minutes.

These ampules were then injected aseptically with 10 μL of the stock solution of TCA in methanol and flame-sealed (Model 524PS manufactured by O. I. Corporation). Final concentrations were approximately 1–3 mg/L. Approximately 0.5 to 1 mL of air space was present in the ampules after sealing.

Ampules were incubated at 28°C (Precision Scientific Model 6) and at 37°C (Precision Scientific Model 4). Experiments at higher temperatures were performed in a Magna-Whirl constant-temperature water bath (Blue M).

Samples were analyzed using a purge and trap device (Tekmar LSC-2), interfaced with a Perkin Elmer Model 8410 GC with FID detector, which employed a 30-m DB-1, 0.53-mm-wide bore capillary column with a 3-μm stationary film thickness (J & W Scientific). The temperature program included a 10-minute hold time at 30°C and a temperature ramp of 5°C/min to 80°C. Helium carrier flow was 2.5 mL/min. Selected analyses were performed by gas chromatography/mass spectroscopy to confirm the formation of 1,1-DCE.

RESULTS

Degradation of TCA

Degradation experiments were performed at various temperatures and sample matrices. The results of these experiments are summarized in Table 2. First-order degradation was observed under the conditions of the experiment and was verified by plotting ln (TCA) versus time. Linear regression analyses were performed on each data set. All rate constants were based on reactions showing a minimum of 75% degradation of the TCA.

Statistical analyses were performed to assess if the slopes measured at any given temperature were significantly different, thus determining the extent to which the sample matrix or pH affects the rate constant. The reaction rates in the buffer solutions (pH 4.5, 7, and 8.5) were not significantly affected by pH ($p < 0.01$). In addition, the rates measured in groundwater matrices

Table 2. Summary of TCA Degradation Rates and Product Formation

Temp. (°C)	Matrix	$10^8 k$ s^{-1}	k_e/k_t (%)	No. Obs.
70	pH 4	1390 ± 85	26 ± 1	9
	pH 5	1530 ± 90		6[a]
	pH 7	1410 ± 100	25 ± 1	8
	pH 10	1400 ± 95	26 ± 2	6
	GW1[b]	1480 ± 90		8[a]
	GW2[b]	1400 ± 80		8[a]
62	pH 13	565 ± 35	38 ± 1	15
53	pH 4.5	140 ± 12	25 ± 2	21[a]
	pH 7.0	140 ± 15	24 ± 2	29[a]
	pH 7.0	144 ± 20	24 ± 2	25
	pH 8.5	145 ± 16	25 ± 2	21[a]
	Seawater	155 ± 18	~25	24
	DW[c]	133 ± 14	23 ± 3	20
39	pH 4.5	25 ± 1.2	19 ± 1	20
	pH 7.0	24 ± 1.1	22 ± 1	14
	pH 8.5	24 ± 1.2	17 ± 1	18
28	pH 4.5	4.4 ± 0.2	23 ± 2	22
	pH 7.0	3.9 ± 0.2	19 ± 2	25
	pH 8.5	4.2 ± 0.2	21 ± 2	23

[a]Cline et al.[9]
[b]Monitoring wells, Orlando, FL.
[c]Distilled deionized water.

at 70°C were not significantly different from rates measured in the buffer solutions.

The spiking solutions were typically prepared with methanol, which resulted in approximately 0.1% methanol in the final solution. Separate experiments were conducted without the use of methanol with no apparent effect on the rate. The use of methanol decreased the variability in concentrations observed among ampules, apparently due to the decreased volatility of TCA in the methanol spiking solution.

Reaction rates in seawater, distilled deionized water, and $0.05\,M$ phosphate buffer solutions showed that the ionic matrix affected the rate of reaction. The fastest rate was observed for seawater, while the rate in distilled deionized water was 14% lower and that in the buffer solutions was approximately 10% lower. Statistically, the differences between the distilled deionized water and the buffer solutions were not significantly different; however, the seawater matrix rate was higher than these at the $p < 0.01$ level. The 10–14% increase in reaction rate observed in the seawater matrix at this temperature may be due to catalysis by some component of that matrix or to the increase of ionized species in the solution.

An Arrhenius plot of the data from this and other studies (Figure 2)

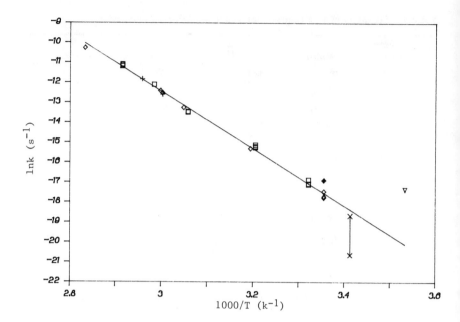

Figure 2.   Arrhenius plot for the degradation of 1,1,1-TCA (□). Additional data from: ◇ Haag and Mill;[11] ◆ Dilling et al.;[10] + Walraevens et al.;[8] x Vogel and McCarty;[10] ▽ Pearson and McConnell.[7]

illustrates the relation between the rate constant and temperature, and includes rates for a variety of matrices including seawater and sodium hydroxide solutions. Since two products were formed, the degradation process was complex, but the overall linearity of the Arrhenius plot implies that a single rate-determining step is involved in the degradation. Based on our results, an activation energy of 119 ± 3 kJ/mol and an A factor of 2.0×10^{13} s^{-1} were calculated. Extrapolated rate constants and estimated half-lives are shown in Table 3.

Included in the Arrhenius plot are the degradation rate for TCA in a pH 13 buffer and the rates calculated by Walraevens et al.[8] for the sodium hydroxide solutions. The rates for these high pH solutions lie within the confidence interval for the regression line. This indicates that the reaction rate was not accelerated by base. The lack of change in the rate in the presence of a high concentration of a strong nucleophile indicates the reaction is primarily SN1, that is, the reaction with the nucleophile occurs after the rate-determining step. The degradation rate is therefore determined by the general ionizing power of the water rather than its nucleophilicity.

The rate data that exceeded the confidence interval of the regression line were from studies[6,10] that estimated the rates of the slow reactions with less than 50% degradation of the parent compound occurring. This procedure

Table 3. Extrapolated Half-Lives for Degradation of TCA

Temp (°C)	Time
15	4.5 years
20	2.0 years
25	10.2 months

generally decreases the correlation coefficient for the rate estimate. The strong relation between temperature and rate observed between 25 and 80°C regardless of sample matrix suggests that rates at temperatures below 25°C can be estimated by extrapolation.

Formation and Stability of 1,1-DCE

The stability of 1,1-DCE was evaluated in concurrent experiments. Preliminary data for the long-term degradation studies at lower temperatures suggested possible degradation of 1,1-DCE;[9] however, after more than one year of incubation no significant degradation was demonstrated at either 28°C or 39°C. In the buffer solutions, seawater, and distilled deionized water, no significant degradation of 1,1-DCE occurred during the course of the evaluation of the degradation of TCA at elevated temperatures. Degradation of 1,1-DCE was observed at pH 13 at a rate of approximately one tenth of the rate of the TCA degradation. An intermediate in the base-catalyzed degradation of 1,1-DCE at elevated temperatures was shown by GC/MS to be chloroacetylene.

Since the degradation of 1,1-DCE was slow compared with the degradation of TCA, the ratio of the rate for elimination (k_e) to the total rate of degradation (k_t) was estimated by plotting the concentration of 1,1-DCE versus $(1 - e^{-kT})$, where T is time. The slope of the line equals ($[TCA]_0 \times (k_e/k_t)$), where $[TCA]_0$ is the concentration of TCA at time 0. The plot for the formation of 1,1-DCE in pH 4 buffer at 28°C is shown in Figure 3.

Increases in pH and/or temperature theoretically favor elimination over substitution.[12] The elimination pathway, according to our data, ranged from 17 to 38% of the total degradation rate of TCA. Higher temperatures showed a slightly larger amount of transformation to 1,1-DCE over the temperature range evaluated in these experiments. The percent elimination was not affected by matrix in the pH range of 4.5 to 8.5. However, the highest percent elimination was measured in the sodium hydroxide (pH 13) solution. Qualitative observations of degradation at approximately 60°C in 0.5, 1.0, and 2.0 molar sodium hydroxide solutions show the presence of 1,1-DCE, and separate experiments indicate that 1,1-DCE also degrades under those conditions. This may account for the lack of detection of 1,1-DCE in experiments by

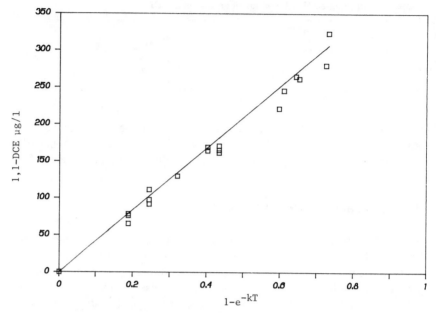

Figure 3. Formation of 1,1-dichloroethene from 1,1,1-TCA in pH 4 buffer at 28°C. When converted to molar concentrations, the slope equals $k_e/k_t \times [TCA]_o$, where k_e is the rate of the elimination pathway, k_t is the total rate of degradation of TCA and $[TCA]_o$ is the initial concentration of TCA.

Walraevens et al.[8] Seawater had no apparent effect on the relative proportion of products.

SUMMARY AND CONCLUSIONS

The abiotic degradation of TCA in water occurs relatively rapidly compared to trends observed for many other alkyl halides. The data obtained in our experiments are consistent with degradation patterns predicted for a true S_N1 or E1 mechanism, with the role of the solvent more related to its overall ionizing power than to its potential role as a nucleophile. The mechanism would involve the formation of an ionic intermediate. This mechanism has been demonstrated for compounds like tertiary butyl chloride.[12] Most primary and secondary alkyl halides follow an S_N2 mechanism, reacting with water as a nucleophile in the intermediate transition state. Chloroform is very resistant to hydrolysis, indicating that steric hindrance may prevent attack by a nucleophile. The stability of an ionic intermediate is surprising, although formation of a carbonium ion can be facilitated by the presence of other halogens on the carbon center.[13]

Defining the mechanism determines the factors which are likely to affect the rate and pattern of the degradation. The primary evidence for the formation of an intermediate that does not involve direct attack by a nucleophile is the lack of increase in the rate constant with the concentration of hydroxide as high as $2.0\,M$. Therefore, high pH or concentrations of nucleophiles would not be expected to accelerate the degradation rate.

The rate-determining step in the degradation of TCA showed minor increases (10–14%) as ionic strength increased from distilled deionized water to seawater matrices, supporting the hypothesis for the formation of an ionic intermediate. A large increase in rate or percentage of elimination in seawater was not demonstrated. Base catalysis resulting from the presence of anions in the buffer solutions was not a factor in the estimate of the degradation rate, although an increase in the elimination pathway occurred at pH 13. Theoretically, addition of organic compounds that would decrease the general ionizing power of the solvent would decrease the rate when compared to that in water.

The activation energy for the rate-determining step as measured by the Arrhenius plot does not separate the activation energies of the elimination and substitution pathways. The extent of transformation to 1,1-DCE was substantially increased at pH 13, while there was negligible effect on the overall rate of degradation. In contrast to the results of Walraevens et al.,[8] 1,1-DCE was detected at elevated hydroxide concentrations. However, degradation of the 1,1-DCE was also observed under those reaction conditions.

The transformation of TCA to 1,1-DCE and the stability of the ethene under typical environmental conditions contribute to the frequency at which this elimination product is observed in water supplies. Although anaerobic biodegradation can occur rapidly under laboratory conditions,[14] field sites in Florida and Arizona demonstrate 1,1-DCE is frequently the dominant degradation product where TCA is present. The major factor affecting the rate of abiotic transformation appears to be temperature.

ACKNOWLEDGMENT

This research was supported by a contract from the Florida Department of Environmental Regulation, Tallahassee, Florida. The project manager was Dr. Geoffery Watts, who provided timely suggestions.

REFERENCES

1. Westrick, J. J., J. W. Mello, and R. F. Thomas. "The Groundwater Supply Survey," *J. Am. Water Works Assoc.* 76:52–59 (1984).

2. "Health Assessment Document for Vinylidene Chloride," U.S. EPA Report-600/8–83/031F (1985).
3. Graf, C. G. "VOC's in Arizona's Groundwater; A Status Report," paper presented at the NWWA FOCUS Conference on Southwestern Ground Water Issues, Tempe, AZ, September 1986.
4. Parsons, F., and G. B. Lage. "Chlorinated Organics in Simulated Groundwater Environments," *J. Am. Water Works Assoc.* 77(5):52–59 (1985).
5. Bouwer, E. J., and P. L. McCarty. "Transformations of 1- and 2-Carbon Compounds under Methanogenic Conditions," *Applied Environ. Microbiol.* 45:1286–1294 (1983).
6. Dilling, W. L., N. B. Tefertiller, and G. J. Kallos. "Evaporation Rates and Reactivities of Methylene Chloride, Chloroform, 1,1,1-Trichloroethane, Trichloroethylene, Tetrachloroethylene, and Other Chlorinated Compounds in Dilute Aqueous Solutions," *Environ. Sci. Technol.* 9:833–838 (1975).
7. Pearson, C. R., and G. McConnell. "Chlorinated C1 and C2 Hydrocarbons in the Marine Environment," *Proc. Roy. Soc. Ser. B* 189:305–332 (1975).
8. Walraevens, R., P. Trouillet, and A. Devos. "Basic Elimination of HCl from Chlorinated Ethanes," *Int. J. Chem. Kinetics* 6:777–786 (1974).
9. Cline, P. V., J. J. Delfino, and W. J. Cooper. "Hydrolysis of 1,1,1-Trichloroethane; Formation of 1,1-Dichloroethene," in *Proceedings of NWWA/API Conference on Petroleum Hydrocarbons and Organic Chemicals in Ground Water: Prevention, Detection and Restoration* (Dublin, OH: National Water Well Association, 1986), 239–247.
10. Vogel, T. M., and P. L. McCarty. "Rate of Abiotic Formation of 1,1-Dichloroethylene from 1,1,1-Trichloroethane in Groundwater," *J. Contaminant Hydrol.* 1:299–308 (1987).
11. Haag, W. R., and T. Mill. "Effect of Subsurface Sediments on Hydrolysis of Haloalkanes and Epoxides," submitted to *Environ. Sci. Technol.* (1987).
12. March, J. *Advanced Organic Chemistry: Reactions, Mechanisms, and Structure* (New York: McGraw-Hill Book Company, 1968), 745–747.
13. Noller, H., and W. Kladnig. "Elimination Reactions over Polar Catalysts: Mechanistic Considerations," *Catal. Rev. Sci. Eng.* 13(2):149–207 (1976).
14. Vogel, T. M., and P. L. McCarty. "Abiotic and Biotic Transformations of 1,1,1-Trichloroethane under Methanogenic Conditions," *Environ. Sci. Technol.* 21:1208–1213 (1987).

SECTION III

Pathogens in Treated Drinking Water

Elimination of Viruses and Bacteria During Drinking Water Treatment: Review of 10 Years of Data from the Montreal Metropolitan Area

Pierre Payment, Centre de Recherche en Virologie, Institut Armand-Frappier, Université du Québec, Laval, Québec, Canada

INTRODUCTION

Pathogenic microorganisms, found in water to be utilized for the preparation of drinking water, should ideally be completely removed or inactivated by the treatment processes applied at the water filtration plant. However, many bacteria, viruses, and parasites have been found to be resistant to one or more of these treatments. The detection of viruses in drinking water meeting current bacteriological standards of quality is a rare occurrence, but since the advent of more reliable methods for their detection the number of reports describing their presence has been increasing. Our own interest in the dissemination and survival of human and animal enteric viruses in water has led us to study not only their presence in surface water, but also their survival during drinking water treatment as well as the health risk they may constitute. The present paper is a review of the results and experience accumulated in our laboratory since 1975.

METHODS

The methods used in our laboratory for the concentration of viruses in water have not been modified much since 1976. The water to be tested is conditioned to pH 3.5 and 0.0015 M aluminum chloride to enhance virus adsorption to the electronegative cartridge filters. After filtration of at least

1000 liters of water, the filters are eluted using an alkaline beef extract solution (1.5%, pH 9.75) that is then flocculated at pH 3.5 to obtain a final volume of concentrate of less than 50 ml, easily assayable in cell culture. The major differences over the years in the assay of these samples has been the use of more sensitive methods, increasing the number of virus types that can be detected. More susceptible cell lines as well as new assay methods have greatly enhanced the overall sensitivity. Our current preferred assay is the use of an immunoperoxidase method with Buffalo Green Monkey kidney cells (BGM) or MA-104 Rhesus kidney cells, which is up to 50 times more sensitive than the previously used cytopathogenic effect method on the same cell lines.

PRESENTATION OF RESULTS

In 1981, we published data showing that viruses were present in all raw and treated drinking water samples tested over a one-year period at a local drinking water treatment plant (plant PV). The plant was using a complete conventional treatment, including prechlorination, flocculation with alum, dynamic sedimentation, slow sand filtration, ozonation, and a final chlorination. At the time, such reports were rare and critiques were rapidly aimed at such causes as laboratory contamination of our samples. This plant was treating water abstracted from a river heavily polluted by untreated sewage discharges and, as discovered over a period of years, was not always properly operated. These poor operation procedures were compounded by an aging plant: flocculation was not always optimal, chlorination levels were not carefully monitored, the ozone generators were not functioning properly, and treatment basins had dead ends, resulting in under treatment. All these reasons were probably sufficient to explain the presence of viruses in the drinking water prepared from river water containing up to several thousand viruses per liter. However, because some treatments such as prechlorination and flocculation reduce dramatically the number of bacteria in water, this plant was still able to produce water meeting current bacteriological standards.

To demonstrate if similar observations could be made at other plants, a Canadian collaborative study among three laboratories was initiated with the financial support of Health and Welfare Canada. Nine drinking water treatment plants, located in the cities of Ottawa, Toronto, and Montreal, were sampled for two years: raw (100 liters) and finished water (1000 liters) were sampled monthly. No viruses were found, but the sensitivity of the cell line used for virus isolation was later found to be very low.

During the same period, we initiated a collaborative study with the Ministry of Environment of Quebec, and in 1982–83 we sampled seven drinking water treatment plants twice monthly for a year. At each plant, 100 to 1000 liters of water were sampled after each treatment: raw water, prechlorination (if possible), sedimentation, filtration, ozonation, and after the final chlorination. Numerous bacteriological and physicochemical analyses were performed. Essentially, all bacteria present in the abstracted water were eliminated by the treatments and, more specifically, by prechlorination and flocculation, which will reduce to barely detectable levels most indicator microorganisms. Bacteria from the total coliform group, as well as some *Pseudomonas aeruginosa,* are occasionally detected in finished water, but they are mostly below the limit of detection of the standard 100 ml test. These low levels of bacterial contaminants are not likely to be of public health significance because the minimal infective dose (the number of bacteria required to initiate the infection) is usually high.

When these waters were tested for the presence of human enteric viruses, the results were quite different (Tables 1 and 2). While the number of infectious viral particles was rarely above 100 viral particles per liter in the river water used, we were still able to detect viruses in finished water at levels of 0.003 to 0.020/liter (3 to 20 viruses/1000 liters).

Table 1.   Virus Elimination During Drinking Water Treatment

Sample	Positive/Total	Virus Density[a]	Residual Virus (%)
Raw	120/152	3.36	100
Chlorinated	11/17	0.072	2.1
Sedimented	23/119	0.016	0.47
Filtered	17/119	0.001	0.03
Ozonated	4/45	0.0003	0.009
Finished	12/138	0.0006	0.018

[a]Virus density expressed as the most probable number of cytopathogenic units per liter (mpncu/L) and as the average of all samples.

Table 2.   Viruses Isolated During Drinking Water Treatments at Water Filtration Plants

Sample	Polio 2	Polio 3	Coxsackie B4	Coxsackie B5
Chlorinated	0	0	0	1
Sedimented	2	6	2	1
Filtered	1	7	3	2
Ozonated	2	2	0	0
Finished	0	9	0	2

The presence of these viruses in drinking water or at any step of treatment did not correlate with the presence of any other bacterial parameter or any physicochemical parameter: the only positive correlation was that if viruses were present at detectable levels in the raw water, they could be detected in the finished water (Table 3).

Thus, our conclusion at the time was: in order to obtain virus-free drinking water, raw water as clean as possible should be used. More recent virological analysis of drinking water at two plants has not revealed the presence of any

Table 3.   Correlation Analysis[a] of Virus Density with Bacteriological Data

Water	Plate Count	Total Coliforms	Fecal Coliforms	*P. aeruginosa*
Raw	0.56	0.68	0.59	0.41
Chlorinated	0.34	0.07	0.04	0.05
Sedimented	0.19	0.03	0.001	0.06
Filtered	0.09	0.09	0.12	0.20
Ozonated	0.11	−0.06	NA[b]	−0.05
Finished	0.06	−0.04	NA[b]	−0.02

[a]Correlation analysis of parameters by Pearson test.
[b]NA = not applicable.

virus; however, these two plants, which had been shown to produce virus-positive drinking water earlier, have been physically modified (new filters, better ozone generators, new settling tanks, etc.) or the operating procedures have been optimized (better control of chlorination and flocculation).

The presence of viruses in the finished water, even after complete treatment, could be explained by two hypotheses: they were resistant to disinfection by chlorine or they were protected by particulate matter. We tested all virus isolates that were detected in treated waters for their resistance to free chlorine. Their resistance to 0.5 mg/L free residual chlorine was highly different (Figure 1). Coxsackievirus B-5 isolates were the most resistant: even after two hours in the presence of residual chlorine more than 20% of the original virus was still infectious. For a similar period of treatment, all poliovirus isolates were reduced to barely detectable levels.

Their presence in drinking water had to be explained by another hypothesis. To determine the physical state of viruses in contaminated river water, we have used a filtration method that has shown that most viruses are free or harbored by particulates less than 0.2 microns in diameter (Figure 2).

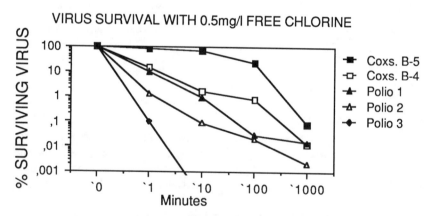

Figure 1.   Inactivation by free residual chlorine (0.4 mg/L) of viruses isolated from treated drinking waters. While poliovirus strains are reduced by more than 99.9% in less than two hours, coxsackievirus strains are much more resistant.

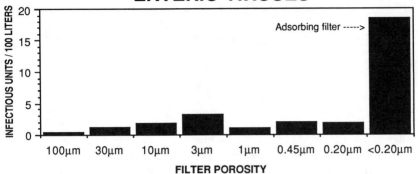

Figure 2. Distribution of human enteric viruses on non-virus-adsorbing filters of selected porosities and on a virus-adsorbent filter. The presence of a large number of viruses on the virus-adsorbing filter indicates that these viruses are embedded or adsorbed to particles that are less than 0.2 μm in diameter.

Because most viruses have diameters that are less than 0.1 micron, these viruses are probably free in the water. The presence of viruses in treated water thus remains to be explained.

As the minimal infective dose of viruses is near 1 infectious particle, the risk of human infection is probably more elevated than for bacteria, for which the number of cells required for inducing infection is more elevated. Dr. C. P. Gerba (University of Arizona), has prepared for the EPA/AAAS a report on the possible health effects of these low doses of viruses present in water.[1] Table 4 summarizes the possible risks estimated by Dr. Gerba according to the viral concentration in water and the infection rate of the virus.

From this table, it is evident that even very low levels of viruses could be the cause of an increased incidence of enteric viral illnesses. The average incidence of gastrointestinal illnesses in the North American population is about 50 episodes/week/1000 individuals, with 1.5 episodes/person/year. For a population of 190,000 individuals in the above example, this is equivalent to about 500,000 episodes per year. These values indicate that the fraction of illnesses due to viral contamination of water at a level of 1 virus/100 liters is low for rate of infection of 1% but can be high for high attack rates. Further analysis of the effects of these low virus levels remains to be evaluated; it is in this direction that our laboratory is now heading.

With the support of the U.S. EPA and of Health and Welfare Canada, we are hoping to initiate an epidemiological surveillance of several hundred families to determine if any health effect is attributable to the presence of these viruses. Preliminary data were obtained during a pilot project to evaluate the feasibility of such a study. Some of the results are presented in Tables

Table 4. Expected Infection Incidence in a Population with 190,000 Individuals at Two Attack Rates

	Infections/Year	
Virus Density	Attack Rate 1%	Attack Rate 30%
1/10 liter	140,000	4,000,000
1/100 liter	14,000	400,000
1/1000 liter	1,400	40,000
1/2000 liter	730	20,000

5 and 6. Data obtained by telephone interviews include water perception in two areas during spring and autumn 1986 as well as the incidence of gastrointestinal symptoms in the spring period of 1986 for bottled water and tap water. From these results, it would appear that the effect of drinking water on the incidence of gastrointestinal illnesses in our area is small: bottled water drinkers have only a slightly reduced incidence of illness. This is, however, in agreement with the hypothesis that only about 5% of these illnesses would be attributed to water.

CONCLUSION

After ten years of experience in the field of environmental virology and particularly in water treatment virology, we have gained some knowledge of the behavior of human enteric viruses, not under laboratory conditions but directly at the water treatment plant. As expected, the theories elaborated from laboratory scale experiments do not always correlate with the experimental data. Viruses were detected after treatments that should theoretically have eliminated more than 12 log of viruses. The data obtained at several water treatment plants have convinced us that, except under optimal conditions, most filtration plants will not remove all human enteric viruses. The low-level viral contamination observed may not be a health problem, but

Table 5. Water Quality Perception in Three Different Cities in Spring and Autumn 1986

	Spring			Autumn		
	Rep.[a] (%)	T-M[b] (%)	Laval (%)	Rep. (%)	T-M (%)	Laval (%)
Excellent	2	4	5	5	8	8
Very good	4	15	19	14	24	23
Good	16	47	48	38	46	47
Poor	23	20	19	31	17	17
Very poor	54	13	8	12	5	5

[a]Rep. = City of Repentigny.
[b]T-M = Terrebonne-Mascouche.

Table 6. Two-week Incidence of Gastrointestinal Symptoms (Spring 1986)

Symptom	Repentigny		Terrebonne	Laval
	Tap Water (%) (N = 626)	Bottled Water (%) (N = 402)	Tap Water (%) (N = 863)	Tap Water (%) (N = 813)
Diarrhea	6.9	6.8	5.6	6.0
Nausea and vomiting	3.6[a]	2.5[a]	1.6	1.2
Nausea and cramps	7.0[a]	5.7[a]	4.3	3.4
Vomiting and cramps	3.1[a]	2.0[a]	1.3	1.2
All symptoms	11.4[a]	9.3[a]	7.5	8.1

[a]Significantly different.

only further studies will establish if such problems exist. Preliminary epidemiological data have enabled us to show that the increased incidence due to drinking water is small and will require the surveillance of large populations to demonstrate any effect. Until then, because these viruses are so resistant to actual water treatment, they can be used as indicators of appropriate water treatment to detect deficiencies.

REFERENCE

1. Gerba, C. P. *Strategies for the Control of Viruses in Drinking Water* (Washington, DC: American Association for the Advancement of Science, 1985), 85 pp.

SELECTED LITERATURE

Payment, P., and M. Trudel. "Detection and Health Risk Associated with Low Virus Concentration in Drinking Water," *Water Sci. Technol.* 17:97–103 (1985).

Payment, P., and M. Trudel. "Immunoperoxidase Method with Human Immune Serum Globulin for Broad Spectrum Detection of Cultivable Viruses: Application to Enumeration of Cultivable Viruses in Environmental Samples," *Appl. Environ. Microbiol.* 50:1308–1310 (1985).

Payment, P., and M. Trudel. "Influence of Inoculum Size, Incubation Temperature and Cell Density on Virus Detection in Environmental Samples," *Can. J. Microbiol.* 31:977–980 (1985).

Payment, P., M. Trudel, and R. Plante. "Elimination of Viruses and Indicator Bacteria at Each Step of Treatment During Preparation of Drinking Water at Seven Water Treatment Plants," *Appl. Environ. Microbiol.* 49:1418–1428 (1985).

Payment, P., M. Trudel, S. A. Sattar, V. S. Springthorpe, T. P. Subrahmanyan, B. E. Gregory, et al. "Virological Examination of Drinking Water: A Canadian Collaborative Study," *Can. J. Microbiol.* 30:105–112 (1984).

CHAPTER 6

Factors Affecting the Occurrence of the Legionnaires' Disease Bacterium in Public Drinking Water Supplies

Stanley J. States, John M. Kuchta, Louis F. Conley, Randy S. Wolford, City of Pittsburgh Water Department, Pittsburgh, Pennsylvania

Robert M. Wadowsky, School of Medicine, University of Pittsburgh, Pittsburgh, Pennsylvania

Robert B. Yee, Graduate School of Public Health, University of Pittsburgh, Pittsburgh, Pennsylvania

INTRODUCTION

Following the initial description of legionnaires' disease and the isolation of the causative agent, *Legionella pneumophila,* it soon became apparent that this bacterium is a common inhabitant of aquatic systems. It is frequently present in natural bodies of water, including those used as the raw source of municipal water supplies.[1,2] It has also been detected, often in even higher numbers, within the internal plumbing systems of hospitals, homes, and other buildings,[3-5] and in bacterial amplifiers such as hot water heaters and cooling towers.[6-8] In some cases, the occurrence of *Legionella* has been associated with disease, while in other cases it has not.[6,9]

The discovery of legionellae in these locations and its demonstrated ability to multiply in tap water[10] raise questions concerning the potential role of public water supplies in contaminating internal plumbing systems. It also suggests a need for a better understanding of environmental factors within drinking water systems that influence the survival and growth of the organism. A series of studies were conducted to investigate these questions. This is a summary of these studies.

SUSCEPTIBILITY OF *L. PNEUMOPHILA* TO CHLORINE

Legionella in a public water supply would be exposed to chlorine concentrations that had been adjusted to control the presence of indicator coliform bacteria. A question arises concerning the susceptibility of legionellae to these levels of disinfectant. A series of experiments were conducted to examine the effectiveness of free chlorine in inactivating *L. pneumophila* at levels typical of municipal drinking water distribution systems.[11,12]

The susceptibility of *L. pneumophila* to chlorine was examined by inoculating tap water with known quantities of legionellae, to achieve a cell density of approximately 3,000 CFU/mL, and treating these aquatic test systems with free chlorine. Viable counts of the bacterium were obtained by plating on buffered charcoal-yeast extract agar. A number of environmental and clinical *L. pneumophila* strains were examined for susceptibility to chlorine disinfection. These included isolates that had been passaged on agar media, as well as non-agar-passaged tap water-maintained strains and legionellae obtained from a hospital plumbing system that had been exposed to relatively high chlorine concentrations from an internal hyperchlorinator. *Legionella* susceptibility to chlorine was compared with that of members of the coliform indicator group. In addition to examining the chlorine susceptibility of legionellae under a set of standard conditions (free Cl_2 concentration = 0.1 mg/L, pH = 7.6, and temperature = 21°C), disinfection experiments were also carried out under other temperatures, pH levels, and chlorine dosages. This was done to simulate the variety of conditions that might be found over time in large municipal treatment and distribution systems.

The results of this study indicate that *Legionella* is more resistant to chlorine than are coliform bacteria. At pH 7.6, temperature = 21°C, and free Cl_2 concentration = 0.1 mg/L, 99% kill of coliforms occurred in less than one minute. This was true for agar-medium-passaged coliforms as well as for a natural population of coliforms in a river water sample. Under the same conditions, 99% inactivation of *L. pneumophila* required 40 minutes. Increasing free chlorine levels to 0.5 mg/L reduced the 99% kill time for *Legionella* to less than five minutes (Figure 1). Decreasing pH also reduced the kill time (Figure 2). Raising the temperature to 32°C decreased inactivation time, while lowering temperature to 4°C increased kill time to greater than 90 minutes (Figure 3). Legionellae strains obtained from hospital plumbing systems that had been continually exposed to higher ambient chlorine concentrations (2 mg/L total chlorine for three weeks) were yet more resistant. Additionally, tap water-adapted, non-agar passaged *L. pneumophila* strains were even more resistant to Cl_2 than agar medium-passaged strains. At pH = 7.6, temperature = 21°C, and free Cl_2 residual = 0.25 mg/L, 99% kill of tap water-maintained strains was achieved within 60 to 90 minutes, compared with 10 minutes for agar-passaged strains (Figure 4).

These studies involved an evaluation of the effectiveness of free chlorine

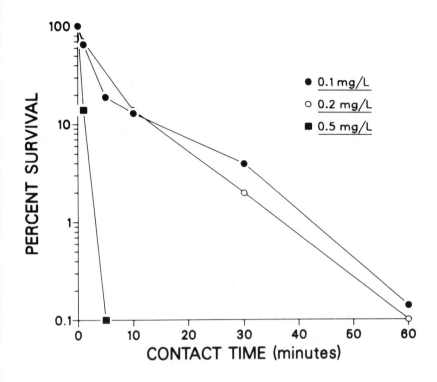

Figure 1. Bactericidal effect of different free chlorine concentrations on agar-grown *L. pneumophila* in tap water at pH 7.6 and 21°C.

in inactivating *Legionella* and an assessment of the coliform bacteria as indicators for this process. The results indicate that legionellae can survive for relatively long periods of time exposed to low concentrations of chlorine. As expected, resistance is enhanced at higher pH levels and lower temperatures, conditions which in public water supplies can occur in different seasons or in different geographic locations. The results also suggest that chlorine resistance may be further enhanced among *Legionella* strains previously exposed to higher ambient chlorine levels as might occur within a municipal drinking water distribution network or a building's internal plumbing system. The observation that more "natural" populations of legionellae are even more resistant to chlorine than their agar-medium-passaged counterparts is consistent with earlier studies on *Pseudomonas aeruginosa*[13,14] and suggests the importance of utilizing non-agar-medium-adapted populations in disinfection studies. Furthermore, the observation that *L. pneumophila* is more resistant to Cl_2 than are coliform bacteria suggests the possibility that small numbers of legionellae may occasionally survive in drinking water that has been judged to be microbiologically acceptable.

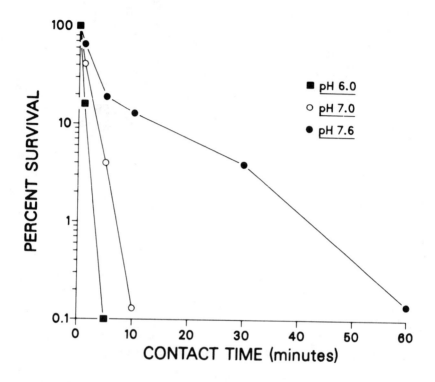

Figure 2.   Effect of pH on bactericidal activity of 0.1 mg/L free chlorine on agar-grown *L. pneumophila* in tap water at 21°C.

EFFECTS OF METALS ON *LEGIONELLA* IN PLUMBING SYSTEMS

The detection of legionellae in tap water and plumbing fixtures in institutional and noninstitutional buildings suggests that this bacterium is able to survive and multiply within internal plumbing systems. Additionally, since the time of the initial isolation of the organism, *L. pneumophila* has been known to have some special metal requirements. Media used for culturing and growing legionellae are routinely supplemented with iron and other metals to enhance recovery and multiplication in the laboratory.[15–18] Given the metallic nature of plumbing systems, a question arises concerning the effects of metals leached from hot water tanks and pipes on survival and growth of *Legionella* species. To gain a better understanding of the ecology of *Legionella* in this habitat, a study was conducted on the chemical environment in plumbing systems and the influence of this metallic environment on the growth of *L. pneumophila*.[19]

The study involved a chemical survey of the environment within hospital hot water plumbing systems. Fifteen water samples, collected from hospital

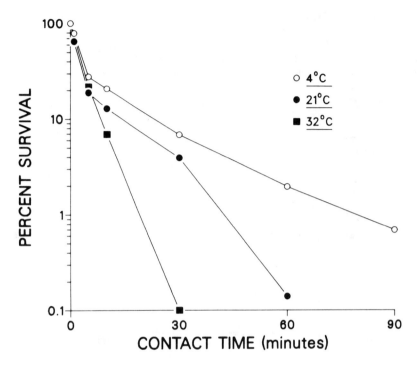

Figure 3. Effect of temperature on bactericidal activity of 0.1 mg/L free chlorine on agar-grown *L. pneumophila* in tap water at pH 7.6.

and institutional hot water tanks known to have supported *L. pneumophila* populations, were analyzed for 23 chemical parameters. The investigation also included two series of growth experiments in which multiplication of legionellae were related to chemical environmental factors. In one set of multiplication experiments, bacterial growth was monitored in tap water samples artificially supplemented with various concentrations of ten metals and inoculated with a non-agar-passaged water stock culture of *L. pneumophila*. The ten metals included: Al, Ca, Cd, Cu, Fe, K, Mg, Mn, Pb, and Zn. In the second set of experiments, bacterial growth was examined in a more "natural" setting by monitoring multiplication in the chemically analyzed water samples, collected as part of the chemical survey of hospital hot water tanks, after sterilization and inoculation with *L. pneumophila*.

Analysis of water samples from institutional hot water tanks indicated that the chemical environment within these tanks varies extensively (Table 1). As a result of corrosion and leaching, the concentrations of certain metals (e.g., Fe and Zn) reach high levels, relative to the low levels of these metals in municipal water entering the building.

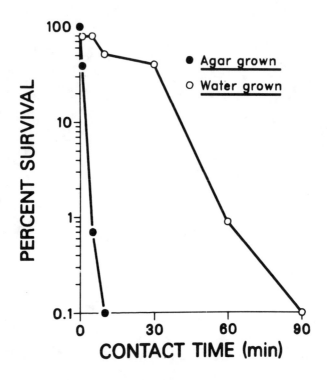

Figure 4. Bactericidal effect of 0.25 mg/L free chlorine on agar-grown and water-grown *L. pneumophila* in tap water at pH 7.6 and 21°C.

In the growth experiments involving metal-supplemented tap water, higher concentrations of all ten metals (10 and 100 mg/L), with the exception of Ca, K, and Mg, produced toxic effects on *L. pneumophila*. Lower concentrations of most of the metals exerted no net influence on *Legionella* populations. However, low levels of "total" Fe and Zn (0.5 and 1.0 mg/L) and higher concentrations of K (1.0, 10, and 100 mg/L) enhanced *L. pneumophila* growth by as much as one log unit over that observed in the unsupplemented tap water control.

In the hot-water tank growth experiment, the 15 samples differed substantially in terms of *L. pneumophila* multiplication. To explain the observed differential growth, a series of linear correlation coefficients were calculated for the association between each of the 23 chemical parameters measured in the hot water tank samples and the greatest population size attained. Fe was the only parameter of the 23 that was significantly correlated ($p = 0.001$). Figure 5 graphically depicts the association between total iron and growth of *L. pneumophila* in the hot-water tank experiment. Growth represents multiplication of legionellae, over a five-week period, from an initial popula-

Table 1.   Chemical Characteristics of 15 Hot-Water Tank Samples[a]

Parameter	Average	Maximum	Minimum
pH	7.62	7.76	7.14
Hardness	97	130	72
Ca	22.6	32.0	12.0
Mg	9.9	19.4	5.4
Alkalinity	32	55	8
Cl	22.0	65.0	14.7
TOC	7.97	86.3	1.42
Fe	12.41	69.97	0.275
Mn	0.646	6.300	0.026
Al	1.98	10.19	0.46
Zn	0.924	7.756	0.020
Cd	0.004	0.018	0.001
Cu	0.469	2.162	0.012
Cr	0.004	0.024	0.001
Pb	0.047	0.178	0.006
Ag	<0.001	0.002	<0.001
Ba	0.082	0.222	0.018
Na	15.20	23.44	11.00
K	2.29	5.50	1.60
As	0.003	0.008	<0.001
Se	0.001	0.007	<0.001
Nl	0.019	0.124	0.001
Co	0.008	0.060	0.001

[a]Concentration = mg/liter.

tion density of approximately 500 CFU/mL. As the histogram indicates, increasing concentrations of iron were associated with enhanced *L. pneumophila* growth until apparently toxic levels were reached. Differences in the absolute concentrations of iron associated with growth enhancement between the metal-supplemented tap water experiments and the hot-water tank experiments may be attributed to matrix differences between the two samples.

This study indicates that while higher levels of most metals are toxic to *Legionella,* lower concentrations of certain metals (e.g., Fe, Zn, and K) may enhance growth. Parallel observations on accompanying non-*Legionellaceae* bacteria failed to show similar growth enhancement. This suggests that metal plumbing components and associated corrosion products are influential in the survival and multiplication of *Legionella* species in plumbing systems.

OCCURRENCE AND POTENTIAL FOR MULTIPLICATION OF *LEGIONELLA* IN MUNICIPAL WATER SYSTEMS

The detection of *Legionella* within the internal plumbing systems of hospitals, private homes, and other buildings has raised the suspicion that this bacterium can survive the drinking water treatment process and that municipal water supplies serve as a pathway for contamination of these build-

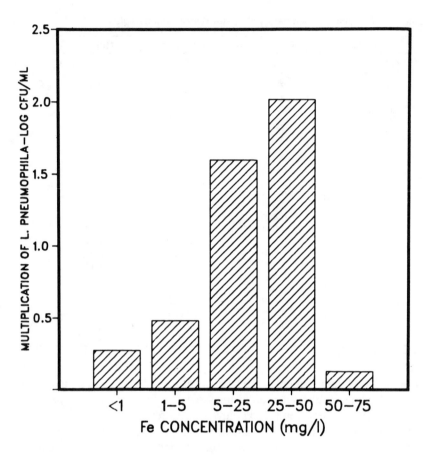

Figure 5.   *L. pneumophila* multiplication and Fe concentration in hot-water tank samples.

ings.[20-23] To investigate this possibility, a study was conducted to determine the occurrence of *Legionella* species in a public water system. An attempt was also made to measure the potential for growth and to determine chemical factors that could influence growth in this system.[24]

Three surveys were conducted, over several years, to isolate *Legionella* from the river water supply, treatment plant, reservoirs, mains, and distribution taps of the City of Pittsburgh water system. A large number of locations were sampled, both on a regularly scheduled basis and randomly. A variety of isolation procedures were utilized, including spread plating, concentration by membrane filtration or flow-through centrifugation, heat/acid enrichment, and direct swabbing of environmental surfaces. A new medium, supplemented with potassium oxalate, shows promise for isolating legionellae from samples that are difficult to culture due to overgrowth.[25] To investigate the potential for *Legionella* growth, additional water samples, collected from throughout the system, were dechlorinated, pasteurized, inoculated with a water stock culture of *L. pneumophila,* and monitored for subsequent bacterial multiplication. The experiment was repeated using water collected during three separate months. Chemical analyses were also performed on these samples to identify factors influencing legionellae multiplication.

The results of the survey for the occurrence of *Legionella* indicated that, despite the use of several isolation techniques, legionellae could not be detected in any of the water system samples other than those from the river or internal hospital plumbing systems. This finding is consistent with earlier studies in which investigators were similarly unsuccessful in isolating viable or virulent *Legionella* sp. from water in the operating portions of treatment plants, finished water reservoirs, or distribution mains.[21,26-28] However, the multiplication experiments revealed a number of habitats within the treatment and distribution system where legionellae could potentially survive and grow (Table 2). These included the bottoms of treatment plant sedimentation basins, the surfaces of treatment plant filters, the bottoms and corners of open, finished water reservoirs, stagnant water collected from the top surface of a vinyl reservoir cover, and samples collected from certain distribution taps. Correlations between chemical characteristics of these water samples and *Legionella* growth potential suggested that turbidity, total organic carbon content, and Cu and Zn concentrations are all positively associated with multiplication.

The inability to isolate legionellae from the municipal treatment and distribution system may be attributed to the fact that in this particular public system a chlorine residual of 0.2 mg/L is generally maintained throughout the distribution system, with even higher residuals detectable within some reservoirs and mains. Alternatively, the failure to detect legionellae could indicate that *Legionella* sp. may only occur within municipal systems sporadically and in low numbers. The multiplication studies revealed a number of locations within the treatment and distribution system where, if the water were dechlorinated, legionellae could multiply. These are locations where water is more stagnant and nutrients are likely to accumulate.

While legionellae were not isolated from this system, the findings indicate that the potential exists for *Legionella* growth within municipal systems. The results also suggest that useful methods to control this contamination include adequate treatment plant filtration; maintenance of a chlorine residual throughout the treatment and distribution network; and effective covering of open reservoirs to eliminate algae, protozoa, and leaf litter that can support *Legionella* multiplication. This study suggests that failure to control *Legionella* growth through these types of measures could result in the seeding of plumbing systems and cooling towers in hospitals and other buildings by legionellae supplied through the municipal water system.

Table 2. *L. pneumophila* **Multiplication After Inoculation in Municipal Water System Samples**

Source	Multiplication (Log CFU/mL) in Month of Sample Collection		
	July	August	October
River	NG	NG	NG
Treatment plant			
Clarifier	NG	1.22	—
Sedimentation basin surface	0.68	NG	NG
Sedimentation basin bottom	—	—	1.34
Sedimentation basin corner	—	—	0.51
Filter surface	—	1.10	NG
Filter effluent	0.07	NG	NG
Clearwell	NG	NG	—
Lanpher Reservoir (open)			
Surface	0.58	1.10	0.16
Bottom	1.93	1.79	0.96
Corner	1.36	0.88	0.11
Brashear Reservoir (open)			
Surface	0.77	1.26	0.12
Bottom	—	1.94	1.08
Corner	0.79	1.24	0.51
McNaugher Reservoir (covered)			
Surface	NG	0.37	—
Bottom	—	1.09	NG
Vinyl cover	0.28	NG	2.02
Hatch	0.89	1.21	NG
Tap samples originating from			
Lanpher Reservoir	0.82	1.08	NG
Brashear Reservoir	1.08	1.03	NG
McNaugher Reservoir	NG	1.71	0.17

ALKALINE TREATMENT OF COOLING TOWERS FOR CONTROL OF *L. PNEUMOPHILA*

Cooling towers and evaporative condensers are often supplied by municipal drinking water systems and have been implicated in a number of outbreaks of legionnaires' disease.[29-33] Surveys of these heat rejection systems, including well-maintained ones, have indicated that bacterial contamination is common and that these systems can serve as amplifiers for *Legionella* species. The use of chlorine and other disinfectants has not been completely effective in controlling this contamination. As a result, it has been suggested that better methods are needed to control legionellae in cooling systems.

Previous field and laboratory studies have indicated that *Legionella* is pH sensitive, multiplying only within the pH range 5.5–9.2.[1,34] Additionally, recent guidelines for the chemical treatment of cooling towers have encouraged maintenance of higher pH levels to facilitate corrosion control and, due to improved chlorine persistence, biofouling control.[35,36]

These factors suggest that operating cooling towers under alkaline conditions may be a suitable method for controlling legionellae in this habitat. Experiments were performed to determine the extent to which pH, alkalinity, and other chemical parameters affect *L. pneumophila* growth in this environment.[37]

Water samples were collected from a hospital cooling tower basin and condenser at two-week intervals over a five-month period. The 13 sets of water samples were analyzed for a variety of chemical parameters. Subsamples of each of the basin and condenser samples were then pasteurized and inoculated with a non-agar-passaged *L. pneumophila* water stock culture obtained from another cooling tower located in the same hospital. These subsamples were evaluated for their ability to support legionellae growth. Additionally, statistical associations were calculated between bacterial growth and the chemical composition of the samples.

The results of the chemical survey indicated that chemical conditions varied widely for almost all of the parameters over the five-month period. Additionally, due to a combination of concentration, treatment, and corrosion effects, the values of most of the chemical factors in the samples differed substantially from those in the make-up tap water. In the multiplication studies, bacterial growth also varied between samples. Statistical analysis indicated that alkalinity and pH were the parameters most closely associated with subsequent legionellae growth. Both parameters were significantly, inversely correlated with the ability of cooling tower water to support *Legionella* growth ($p = 0.005$ to 0.001). The relationship between *L. pneumophila* growth and water sample alkalinity and pH is shown graphically in Figures 6 and 7. Growth is expressed as multiplication of legionellae, over a five-week period, from an initial population density of approximately 1,000 CFU/mL.

These data suggest that elevated pH and alkalinity bring environmental conditions out of the tolerance range of *Legionella* species and would be an additional tool for controlling multiplication in cooling systems. Since current recommendations encourage the use of higher pH for the control of corrosion and fouling, the use of such an approach would be additionally beneficial from a general maintenance viewpoint.

CONCLUSIONS

These studies on the survival, multiplication, and inactivation of *Legionella* in municipal water supplies, internal plumbing systems, and bacterial amplifiers indicate the following conclusions:

1. *Legionella* appears to be less susceptible to chlorine than are the indicator coliform bacteria. The resistance is greater for non-agar-passaged than for agar-medium-passaged strains. Chlorine resistance is further enhanced by higher pH levels, lower temperatures, and previous exposure to chlorine. These results suggest that *Legionella* could potentially survive conventional chlorine levels typically found within portions of municipal drinking water systems.

2. Metal plumbing components and associated corrosion products are important factors in the survival and growth of *L. pneumophila* in plumbing systems. While high levels of certain metals suppress multiplication, lower levels of key metals enhance growth.

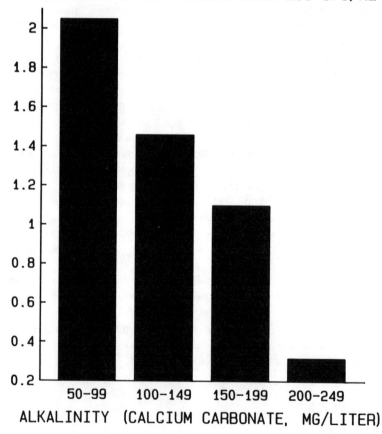

Figure 6. *L. pneumophila* multiplication and alkalinity in cooling tower condenser samples.

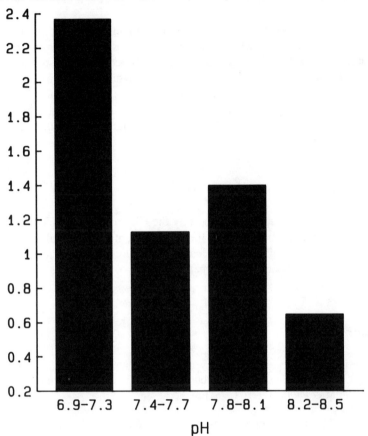

MULTIPLICATION OF L. PNEUMOPHILA—LOG CFU/ML

Figure 7. *L. pneumophila* multiplication and pH in cooling tower condenser samples.

3. Although *Legionella* is not readily isolated from municipal water supplies, the systems are capable of supporting its growth. Prevention of legionellae growth is aided by diligent application of conventional water treatment practices. These include maintenance of chlorine residuals, especially in secluded sites such as the bottoms of reservoirs, effective filtration, and reduction of growth-supporting algae, protozoa, and leaf litter by covering sedimentation basins and reservoirs. Failure to control *Legionella* growth through these types of measures can potentially result in the seeding of plumbing systems and cooling towers in hospitals and other buildings by legionellae supplied through the public water supply.

4. Operating cooling towers outside of the optimal environmental tolerance range for *Legionella* multiplication (e.g., at elevated alkalinities and pH > 9) may be a useful approach to controlling growth in this habitat.

While this work has examined factors affecting the occurrence of the legionnaires' disease bacterium in water supplies, it is important to emphasize that the extent of the hazard posed by this occurrence is still not entirely clear. Legionellae have been isolated from natural and man-made aquatic systems that have been associated with outbreaks of legionnaires' disease as well as in situations in which no disease has been documented. Further study is needed to assess the significance of *Legionella* contamination and factors that might influence this significance.

ACKNOWLEDGMENTS

This work was supported in part by the City of Pittsburgh Water Department. It was also sponsored by the Environmental Epidemiology Center of the Graduate School of Public Health of the University of Pittsburgh under the support of cooperative agreement CR80681–01–2 with the U.S. Environmental Protection Agency.

REFERENCES

1. Fliermans, C. B., W. B. Cherry, L. H. Orrison, S. J. Smith, D. L. Tison, and D. H. Pope. "Ecological Distribution of *Legionella pneumophila*," *Appl. Environ. Microbiol.* 41:9–16 (1981).
2. Ortiz-Rogue, C. M., and T. C. Hazen. "Abundance and Distribution of Legionellaceae in Puerto Rican Waters," *Appl. Environ. Microbiol.* 53:2231–2236 (1987).
3. Tobin, J. O., C. L. R. Bartlett, S. A. Waitkins, G. Macrae, A. G. Taylor, R. J. Fallon, and F. R. N. Lynch. "Legionnaires' Disease: Further Evidence to Implicate Water Storage and Water Distribution Systems as Sources," *Brit. Med. J.* 282:573 (1981).
4. Dennis, P. J., J. A. Taylor, R. B. Fitzgeorge, C. L. R. Bartlett, and G. I. Barrow. "*Legionella pneumophila* in Water Plumbing Systems," *Lancet* i:949–951 (1982).
5. Stout, J. E., V. L. Yu, and P. Muraca. "Legionnaires' Disease Acquired Within the Homes of Two Patients," *J. Am. Med. Assoc.* 257:1215–1217 (1987).
6. Wadowsky, R. M., R. B. Yee, L. Mezmar, E. J. Wing, and J. N. Dowling. "Hot Water Systems as Sources of *Legionella pneumophila* in Hospital and Nonhospital Plumbing Fixtures," *Appl. Environ. Microbiol.* 43:1104–1110 (1982).

7. Fliermans, C. B., G. E. Bettinger, and A. W. Fynsk. "Treatment of Cooling Systems Containing High Levels of *Legionella pneumophila*," *Water Res.* 16:903–909 (1982).

8. Kurtz, J. B., C. L. R. Bartlett, U. A. Newton., R. A. White, and N. L. Jones. "*Legionella pneumophila* in Cooling Water Systems: Report of a Survey of Cooling Towers in London and a Pilot Trial of Selected Biocides," *J. Hyg. Camb.* 88:369–381 (1982).

9. Fliermans, C. B., W. B. Cherry, L. H. Orrison, and L. Thaker. "Isolation of *Legionella pneumophila* from Nonepidemic-Related Aquatic Habitats," *Appl. Environ. Microbiol.* 37:1239–1242 (1979).

10. Yee, R. B., and R. M. Wadowsky. "Multiplication of *Legionella pneumophila* in Unsterilized Tap Water," *Appl. Environ. Microbiol.* 43:1330–1334 (1982).

11. Kuchta, J. M., S. J. States, A. M. McNamara, R. M. Wadowsky, and R. B. Yee. "Susceptibility of *Legionella pneumophila* to Chlorine in Tap Water," *Appl. Environ. Microbiol.* 46:1134–1139 (1983).

12. Kuchta, J. M., S. J. States, J. E. McGlaughlin, J. H. Overmeyer, R. M. Wadowsky, A. M. McNamara, R. S. Wolford, and R. B. Yee. "Enhanced Chlorine Resistance of Tap Water-Adapted *Legionella pneumophila* as Compared with Agar-Medium-Passaged Strains," *Appl. Environ. Microbiol.* 50:21–26 (1985).

13. Favero, J. S., L. A. Carson, W. W. Bond, and N. J. Petersen. "*Pseudomonas aeruginosa:* Growth in Distilled Water from Hospitals," *Science* 173:836–838 (1971).

14. Carson, L. A., M. S. Favero, W. W. Bond, and N. J. Petersen. "Factors Affecting Comparative Resistance of Naturally Occurring and Subcultured *Pseudomonas aeruginosa* to Disinfectants," *Appl. Environ. Microbiol.* 23:863–869 (1972).

15. Feeley, J. C., G. W. Gorman, R. E. Weaver, D. C. Mackel, and H. W. Smith. "Primary Isolation Media for Legionnaires' Disease Bacterium," *J. Clin. Microbiol.* 8:320–325 (1978).

16. Wadowsky, R. M., and R. B. Yee. "Glycine-Containing Selective Medium for Isolation of *Legionellaceae* from Environmental Specimens," *Appl. Environ. Microbiol.* 42:768–772 (1981).

17. Reeves, M. W., L. Pine, S. H. Hutner, J. R. George, and W. K. Harrell. "Metal Requirements of *Legionella pneumophila*," *J. Clin. Microbiol.* 13:688–695 (1981).

18. Tesh, M. J., and R. D. Miller. "Growth of *Legionella pneumophila* in Defined Media: Requirement for Magnesium and Potassium," *Can. J. Microbiol.* 28:1055–1058 (1982).

19. States, S. J., L. F. Conley, M. Ceraso, T. E. Stephenson, R. S. Wolford, R. M. Wadowsky, A. M. McNamara, and R. B. Yee. "Effects of Metals on *Legionella pneumophila* Growth in Drinking Water Plumbing Systems," *Appl. Environ. Microbiol.* 50:1149–1154 (1985).

20. Stout, J., V. L. Yu, R. M. Vickers, J. Zuranleff, M. Best, A. Brown, R. B. Yee, and R. Wadowsky. "Ubiquitousness of *Legionella pneumophila* in the Water Supply of a Hospital with Endemic Legionnaires' Disease," *N. Engl. J. Med.* 306:466–468 (1982).

21. Tison, D. L., and R. J. Seidler. "*Legionella* Incidence and Density in Potable Drinking Water Supplies," *Appl. Environ. Microbiol.* 45:337–339 (1983).
22. Hsu, S. C., R. Martin, and B. B. Wentworth. "Isolation of *Legionella* Species from Drinking Water," *Appl. Environ. Microbiol.* 18:830–832 (1984).
23. Fliermans, C.B. "Philosophical Ecology: *Legionella* in Historical Perspective," in *Legionella: Proceedings of the 2nd International Symposium,* C. Thornsberry, A. Balows, J. C. Feeley, and W. Jakubowski, Eds. (Washington, D.C.: American Society for Microbiology, 1984), 285.
24. States, S. J., L. F. Conley, J. M. Kuchta, B. M. Oleck, M. J. Lipovich, R. S. Wolford, R. M. Wadowsky, A. M. McNamara, J. L. Sykora, G. Keleti, and R. B. Yee. "Survival and Multiplication of *Legionella pneumophila* in Municipal Drinking Water Systems," *Appl. Environ. Microbiol.* 53:979–986 (1987).
25. Paul, M. A., and J. M. Kuchta. "Potassium Oxalate as a Selective Agent in *Legionella pneumophila* Isolation," in *Abstr. Ann. Mtg.* (Washington, DC: American Society for Microbiology, 1986) Q87, 298.
26. Fisher-Hoch, S. P., J. O'H. Tobin, A. M. Nelson, M. G. Smith, J. M. Talbot, C. L. R. Bartlett, M. B. Gilbert, J. E. Pritchard, R. A. Swann, and J. A. Thomas. "Investigation and Control of an Outbreak of Legionnaires' Disease in a District General Hospital," *Lancet* i:932–936 (1981).
27. Voss, L., K. S. Button, R. C. Lorenz, and O. H. Tuovinen. "*Legionella* Contamination of a Preoperational Treatment Plant," *J. Am. Water Works Assoc.* 78:70–75 (1986).
28. Barbaree, J. M., G. W. Gorman, W. T. Martin, B. S. Fields, and W. E. Morrill. "Protocol for Sampling Environmental Sites for Legionellae," *Appl. Environ. Microbiol.* 53:1454–1458 (1987).
29. Glick, T. H., M. B. Gregg, B. Berman, G. F. Mallison, W. W. Rhodes, and I. Kassanoff. "Pontiac Fever: An Epidemic of Unknown Etiology in a Health Department I. Clinical and Epidemiological Aspects," *Am. J. Epidemiol.* 107:149–160 (1978).
30. Cordes, L. G., D. W. Fraser, P. Skaliy, C. A. Perlino, W. R. Elsea, G. F. Mallison, and P. S. Hayes. "Legionnaires' Disease Outbreak at an Atlanta, Georgia Country Club: Evidence for Spread from an Evaporative Condenser," *Am. J. Epidemiol.* 111:425–431 (1980).
31. Dondero, T. J., R. C. Rendtorff, G. F. Mallison, R. M. Weeks, J. S. Levy, E. W. Wong, and W. Schaffner. "An Outbreak of Legionnaires' Disease Associated with a Contaminated Air Conditioning Cooling Tower," *N. Engl. J. Med.* 302:362–370 (1980).
32. Band, J. D., M. LaVenture, J. P. David, G. F. Mallison, P. Skaliy, P. S. Hayes, W. L. Schell, H. Weiss, D. J. Greenberg, and D. W. Fraser. "Endemic Legionnaires' Disease: Airborne Transmission Down a Chimney," *J. Am. Med. Assoc.* 245:2404–2407 (1981).
33. Gorman, G. W., J. C. Feeley, A. Steigerwalt, P. H. Edelstein, C. W. Moss, and D. J. Brenner. "*Legionella anisa:* A New Species of *Legionella* Isolated from Potable Waters and a Cooling Tower," *Appl. Environ. Microbiol.* 49:305–309 (1985).
34. Wadowsky, R. M., R. Wolford, A. M. McNamara, and R. B. Yee. "Effect of Temperature, pH, and Oxygen Level on the Multiplication of Naturally Occur-

ring *Legionella pneumophila* in Potable Water," *Appl. Environ. Microbiol.* 49:1197–1205 (1985).

35. Kemmer, F. N., and J. McCaillion, Eds. *The Nalco Water Handbook* (New York: McGraw-Hill Book Co., 1979).

36. Characklis, W. G., M. G. Trulear, N. Stathopoulos, and L. C. Chang. "Oxidation and Destruction of Microbial Films," in *Water Chlorination: Environmental Impact and Health Effects,* R. L. Jolley, Ed. (Stoneham, MA: Butterworth Publishers, 1980), 349.

37. States, S. J., L. F. Conley, S. G. Towner, R. S. Wolford, T. E. Stephenson, A. M. McNamara, R. M. Wadowsky, and R. B. Yee. "An Alkaline Approach to Treating Cooling Towers for Control of *Legionella pneumophila,*" *Appl. Environ. Microbiol.* 53:1775–1779 (1987).

CHAPTER 7

Presence of Fungi in Drinking Water

William D. Rosenzweig, Department of Biology, West Chester State University, West Chester, Pennsylvania

Wesley O. Pipes, Department of Bioscience and Biotechnology, Drexel University, Philadelphia, Pennsylvania

INTRODUCTION

Fungi are a rather large and diverse group of eucaryotic organisms which are active in degrading organic substrates in most habitats, including those of aquatic environments. Although there have been few studies of fungi in water distribution systems, it is likely that they are active therein. This chapter gives a brief summary of previously published reports and then presents some recent results from our field surveys.

Occurrence in Water Systems

There are only a few papers which report surveys of fungi in distribution systems. In Great Britain, Bays et al.[1] studied a surface water system and found fungi present in service mains. *Cephalosporium, Verticillium, Trichoderma, Nectria, Phoma,* and *Phialophora* were the most common isolates. Nagy and Olson[2] reported that the mean number of fungal colony forming units (CFU) was 18 per 100 mL for an unchlorinated groundwater system and 34 per 100 mL for a chlorinated surface water system. The most commonly isolated genera were *Acremonium, Paecilomyces, Penicillium,* and *Sporocybe.* Nagy and Olson[3] also reported isolating fungi from the surfaces of water mains removed during routine maintenance. Three-fourths of all pipe surfaces examined had filamentous fungi, and half had yeast. Members of the genera *Penicillium, Nomurae, Sporocybe,* and *Acremonium* were the most frequently isolated filamentous forms, while *Cryptococcus* and *Rhodotorula* were the most common yeast isolates.

Rosenzweig et al.[4,5] investigated six chlorinated groundwater systems in Pennsylvania and New Jersey. Over half of their samples had fungi present, with typical counts between 1 and 6 colony forming units (CFU) per 50 mL of water. Niemi et al.[6] found fungi to be present in 29 of 32 treated water samples and 30 of 32 tap water samples from a system where the surface water supply was treated by sand filtration and disinfection. Treatment of the raw water by chemical coagulation and disinfection was effective in removing fungi. West[7] isolated fungi from a water distribution system supplied from an oligotrophic lake in Nevada (Lake Mead). She found an average count of 1.5 CFU per 50 mL and 17.6% of 978 samples were positive for fungi. The most common isolates were *Cladosporium, Phoma, Alternaria, Candida,* and *Rhodotorula.*

Fungi have also been found in water distribution systems in Salem and Beverly, Massachusetts,[8] Riverside, California,[9] Sioux Falls, South Dakota,[10] and the cities of Nancy and Metz in France.[11] The yeast counts in the Sioux Falls system were as high as 1000 per mL.[10] Seidler et al.[12] found fungal mycelia embedded in the microbial slime that accumulates on the surface of redwood water storage tanks used for private water supplies in various locations throughout Oregon. Donlan[13] found fungi among the microorganisms developing on clean iron surfaces in water systems in Pennsylvania.

Treated water may carry fungal spores into water distribution systems, or airborne spores may enter through standpipes and elevated storage tanks. Distribution system storage facilities normally provide an interface where an exchange of air between the tank and the outside atmosphere occurs. If these spores are able to survive and grow in distribution systems, they may have some health or water quality significance.

Possible Significance of Fungi in Water Systems

Chlorine can react with fungal growths in distribution systems, thereby reducing the chlorine residual and the protection which it affords. The chlorine demand of fungal conidia[14] has been measured to be in the range of 3.6×10^{-9} to 3.2×10^{-8} mg per spore, while vegetative yeast cells have a demand between 1.2 and 8.0×10^{-9} mg per cell. The chlorine demand of fungal spores and yeast cells is one to two orders of magnitude greater than that of the cells of coliform bacteria.[14] Since the volume of fungal spores and yeast cells is about three orders of magnitude greater than the volume of bacterial cells and the surface area ratio is only about two orders of magnitude, it appears than the chlorine demand may be more closely related to the surface area rather than the volume of the cells. The reaction of chlorine with fungal spores and mycelia could produce chlorinated organic compounds, some of which might be of health concern.

Slime on the inside of redwood storage tanks was found to contain

Klebsiella pneumoniae and *Enterobacter* sp. embedded with fungal myce-lia.[12] The presence of the fungi may have afforded the bacteria some protec-tion from the chlorine. The presence of the coliform bacteria indicates sew-age contamination at some point. However, survival on fungal mycelia could lead to extended false indications of continuing sewage contamination.

Recent evidence suggests that microhabitats such as pits or cracks on the inner surface of pipes, tubercles, and suspended particles can serve as sites for growth and colonization by bacteria.[15–19] These microhabitats could also provide a place for the growth of fungi, and the fungal mycelia could, in turn, form a substrate for bacterial attachment and growth. It is also possible that some bacteria might be able to digest and metabolize the extracellular material produced by the fungi. Tubercle material on corroded iron pipes has been found to contain humic material and have an organic carbon content of approximately 2%.[18] It has also been demonstrated that when point-of-use water treatment filters, commonly sold for attachment to home water faucets, are used they tend to trap particulate organic material present in the water.[20] These filters then become sites for colonization by microorganisms, includ-ing fungi.

Rosenzweig et al.[4] found 8 strains of *Aspergillus flavus,* 11 strains of the human pathogen *Aspergillus niger,* and 3 isolates of the pathogen *Aspergillus fumigatus.* Roesch and Leong[9] isolated fungi on polyurethane caulking used to seal leaking joints of a water main in Southern California. One of the isolates was a human pathogen, *Petriellidium boydii.* Recent evidence[21,22] suggests that fungi present in potable water can also lead to various allergic reactions in exposed individuals. The use of water with fungi present for the preparation of foods, bathing and showering, mist humidifiers, or drinking could lead to contact between fungal pathogens and the body, which could result in an infection. Also, toxin producers might be introduced into an environment suitable for toxin production.

Fungi may be able to decompose some of the materials used in construc-tion of a distribution system. Jute packing was used in water main joints for many years, but more recently natural or synthetic rubber O-ring gaskets lubricated with a nontoxic grease have been used.[23] Jute is readily attacked by a number of fungi and probably has disappeared from most distribution systems by now. Burman[24] demonstrated that fungi could grow under water, using mastic jointing compounds as the organic substrate. Fungi have also been found to be responsible for the deterioration of a polyurethane jointing compound used in two underground reservoirs in Great Britain.[25] Burman and Colbourne[26] have also reported that various plumbing materials, includ-ing jointing material, plastics, paints, and greases, will support fungal growth.

It has been suggested by Bays et al.[1] and Burman[24] that fungi may be partially responsible for various taste and odor problems that develop in distribution systems. This is especially true if the water is locally warmed and subjected to stagnation or low flow rates for long periods of time.

Objectives of This Investigation

Although the number of reports is small, it is apparent that fungi can be found in water samples from distribution systems in the United States and Europe. Water systems in other parts of the world have not been examined for fungi, and the methods used for isolation have been very limited. It is quite likely that fungi could be found in water systems in other areas and other fungi could be found if the search were more general.

During 1986 and 1987, we have undertaken an extension of our previous studies of the occurrence of fungi in small water distribution systems in the Philadelphia area. These results are reported herein, added to the previous reported results[4,5] and interpreted in terms of the survival of fungi in water systems.

MATERIALS AND METHODS

Distribution Systems Sampled

Five of the distribution systems from which these samples were obtained are groundwater systems that have been described previously.[27] Four of the five systems have no treatment other than disinfection using chlorine. They include the following systems: SR, chlorination dose = 1.0 mg/L; MW, chlorination dose = 1.0 mg/L; BG, chlorination dose = 2.0 mg/L; WH, chlorination dose = 1.4 mg/L; and BL, chlorination dose = 10.0 mg/L. The water for system BL is also aerated for H_2S and CO_2 removal and then filtered through sand. Samples from two surface water systems are included in this analysis. Both of these systems use water from streams that receive runoff from agricultural areas and residential developments where septic tanks are common. Both systems provide treatment consisting of coagulation, settling, filtration, and disinfection with chlorine. System FV postchlorinates at a dose of 1.5 mg/L, and system WC postchlorinates at a dose of 0.8 mg/L.

Sampling Procedures

The systems were sampled at various times during November–December 1980, June 1981, May–October 1986, and April–July 1987. Residential taps, elevated storage tanks, fire hydrants, chlorinated water in clearwells, aerated water, and raw water sources were sampled. Temperature and residual chlorine concentrations were measured when the samples were collected according to the procedures in the fifteenth edition of *Standard Methods*.[28] Samples were collected in sterile bottles to which thiosulfate solution had been added to neutralize any free residual chlorine. Fungi were isolated from 50-mL portions of the samples (1-mL portions in the case of the raw water samples from system WC) by the membrane filter (MF) procedure,[29,30] using 0.45–5 μm filters. The filters were incubated on plates of Sabouraud dextrose agar that had been supplemented with rose bengal (33.3 mg/L) and streptomycin (80.0 mg/L) and were scored daily for two weeks. Isolated fungal colonies were transferred to slants of Sabouraud dextrose agar. The fungi were identified by using various taxonomic guides and monographs.[31–35]

RESULTS AND DISCUSSION

The data representing the occurrence of fungi in various parts of water systems are presented in Tables 1 and 2. Most of the raw (before treatment) water samples had fungi present and the counts per unit volume were relatively high. Treatment, even if it consisted of only chlorination, reduced the percentage of the samples with fungi present and the fungus count in the positive samples. The lowest occurrence and densities of fungi were found in the samples from residential taps in the groundwater systems. The lowest occurrence and densities of fungi in the surface water systems were in the treated water (clearwell). In both cases, the fungal occurrence and densities were higher in the storage tank samples than in the residential tap samples.

The average fungal count per positive sample was highest in the storage tanks for groundwater systems and in fire hydrant samples for surface water systems. The frequency of occurrence of fungi in samples from residential taps is much lower, perhaps indicating some die-off of spores in the distribution system. As previously noted, fungal conidia and yeast cells react with chlorine.[14] Thus, the chlorine doses required to achieve a 99.9% reduction in fungal counts depends upon the initial fungal titer in the water. Chlorine doses in the range of 1 to 2 mg/L free chlorine are probably adequate for inactivation of fungal propagules in water if there is adequate contact time. However, most distribution systems have sediment in the bottom of mains, joints, and/or corrosion tubercles, which provide habitats in which fungi may be protected from the chlorine in the water. Such evidence as is available indicates that many fungal spores and yeast cells survive in water distribution systems. Free chlorine can inactivate fungi in clean systems but it would probably require a major effort at flushing, cleaning, relining, and disinfection to eliminate fungi from many systems.

The genera of fungi which were isolated from the five groundwater systems and the two surface water systems of this study are listed in Table 3. There is no evidence in these lists of differences in fungi present between groundwater and surface water systems. All of the fungi listed are common

TABLE 1.   Fungi from Different Sampling Points of Five Groundwater Systems

Type of Sample	Number of Samples	Percent Positive	Average CFU Per Positive 50-mL Sample
Well water	37	70.3	8.2
Aerated water	3	66.7	1.5
Finished water	7	28.6	2.5
Clearwell	84	50.0	6.6
Residential tap	484	46.1	3.8
Fire hydrants	59	94.4	6.3
Storage tank	53	94.3	10.3
Total	727	55.2	5.9

TABLE 2.   Fungi from Different Parts of Two Surface Water Systems

Type of Sample	Number of Samples	Percent Positive	Average CFU Per Positive Sample[a]
Raw water (stream)	120	86.7	4.7
Clearwell	40	10.0	1.8
Residential tap	648	27.5	4.2
Fire Hydrants	96	63.5	5.6
Storage tank	87	53.3	4.9
Total	991	40.8	4.6

[a]All samples were 50 mL except the raw water samples, which were 1 mL.

TABLE 3.   Genera of Fungi Isolated from Water Distribution Systems

Genus	Five Groundwater Systems	Two Surface Water Systems
Alternaria	X	X
Aspergillus	X	X
Candida (yeast)	X	X
Cephalosporium	X	X
Cladosporium	X	X
Cryptococcus (yeast)	X	X
Cunninghamella	X	
Epicoccum	X	X
Fusarium	X	X
Geotrichum	X	X
Mucor	X	X
Paecilomyces	X	
Penicillium	X	X
Peyronellaea	X	
Phaeococcus (yeast)	X	X
Phialophora	X	X
Phoma	X	X
Pithomyces	X	X
Rhodotorula (yeast)	X	X
Sterile mycelium	X	X
Trichoderma	X	X
Trichosporon		X
Verticillium	X	X

soil organisms, and their spores are found in the atmosphere. Except for *Cunninghamella* and *Mucor* they are all imperfect fungi, and in that regard they are not particularly different from the fungi isolated from water systems by other investigators. These results are few and other, possibly different, results may be expected in the future.

CONCLUSIONS

Apparently some fungi survive water treatment processes in small numbers, although the majority are eliminated. Relatively large numbers of fungi can be isolated from various parts of distribution systems. Included among the known isolates are various human pathogens. Investigations have implicated fungi in the deterioration of water quality and the degradation of some materials in water distribution systems. However, the exact role fungi might play in relation to human health and interference with the operation of the distribution system will require additional studies.

ACKNOWLEDGMENT

This study was supported in part by a grant from the National Science Foundation, Division for Fundamental Research in Emerging and Critical Engineering Systems (ECE-8515710).

REFERENCES

1. Bays, L. R., N. P. Burman, and W. M. Lewis. "Taste and Odour in Water Supplies in Great Britain: A Survey of the Present Position and Problems for the Future," *Water Treatment Exam*. 19:136–160 (1970).
2. Nagy, L. A., and B. H. Olson. "The Occurrence of Filamentous Fungi in Drinking Water Distribution System," *Can. J. Microbiol*. 28:667–671 (1982).
3. Nagy, L. A., and B. H. Olson. "Occurrence and Significance of Bacteria, Fungi, and Yeasts Associated with Distribution Pipe Surfaces," in *Proceedings of the Water Quality Technology Conference* (Denver: American Water Works Association, 1985), 213–238.
4. Rosenzweig, W. D., H. Minnigh, and W. O. Pipes. "Presence of Fungi in Potable Water Distribution Systems," *J. Am. Water Works Assoc*. 78:53–55 (1986).
5. Rosenzweig, W. D., and W. O. Pipes. "Survival of Fungi in Potable Water Systems, in *Proceedings of the Water Quality Technology Conference* (Denver: American Water Works Association, 1987), 449–456.
6. Niemi, R. M., S. Knuth, and K. Lundstrom. "Actinomycetes and Fungi in Surface Waters and in Potable Water," *Appl. Environ. Microbiol*. 43:378–388 (1982).

7. West, P. R. "Isolation Rates and Characterization of Fungi in Drinking Water Distribution Systems," in *Proceedings of the Water Quality Technology Conference* (Denver: American Water Works Association, 1986), 457–473.
8. Reilly, J. K., and J. S. Kippin. "Relationship of Bacterial Counts with Turbidity and Free Chlorine in Two Distribution Systems," *J. Am. Water Works Assoc.* 75:309–312 (1983).
9. Roesch, S. C., and L. Y. C. Leong. "Isolation and Identification of *Petriellidium boydii* from a Municipal Water System," in *Abstr. Ann. Mtg.* (Washington, DC: American Society for Microbiology, 1983), 276.
10. O'Connor, J. T., L. Hash, and A. B. Edwards. "Deterioration of Water Quality in Distribution Systems," *J. Am. Water Works Assoc.* 67:113–116 (1975).
11. Hinzelin, F., and J. C. Block. "Yeasts and Filamentous Fungi in Drinking Water," *Environ. Letters* 6:101–106 (1985).
12. Seidler, R. J., J. E. Morrow, and S. T. Bogley. "*Klebsielleae* in Drinking Water Emanating from Redwood Tanks," *Appl. Environ. Microbiol.* 33:893–900 (1977).
13. Donlan, R. M. "An Investigation of the Biofilm Developing on Cast Iron Surfaces Exposed in Drinking Water Mains," PhD Thesis, The Environmental Studies Institute, Drexel University, Philadelphia (1987).
14. Rosenzweig, W. D., H. A. Minnigh, and W. O. Pipes. "Chlorine Demand and Inactivation of Fungal Propagules," *Appl. Environ. Microbiol.* 45:182–186 (1983).
15. Ridgway, J. F., and B. H. Olson. "Scanning Electron Microscope Evidence for Bacterial Colonization of a Drinking-Water Distribution System," *Appl. Environ. Microbiol.* 41:274–287 (1981).
16. Cundell, A. M., and A. P. Mulcock. "The Biodeterioration of Natural Rubber Pipe-Joint Rings in Sewer Mains," in *Proceedings of the Third International Biodegradation Symposium*, J. M. Sharpley and A. M. Kaplan, Eds. (London: Applied Science Publishers, 1976), 659–664.
17. Kennedy, H. "External Loads and Foundations for Pipes," *J. Am. Water Works Assoc.* 63:189–196 (1971).
18. Tuovinen, O. H., K. S. Button, A. Vuorinen, L. Carlson, D. M. Mair, and L. A. Yut. "Bacterial, Chemical, and Mineralogical Characteristics of Tubercles in Distribution Pipelines," *J. Am. Water Works Assoc.* 72:626–635 (1980).
19. Tuovinen, O. H., and J. C. Hsu. "Aerobic and Anaerobic Microorganisms in Tubercles of the Columbus, Ohio, Water Distribution System," *Appl. Environ. Microbiol.* 44:761–764 (1982).
20. Tobin, R. S., D. K. Smith, and J. A. Lindsay. "Effects of Activated Carbon and Bacteriostatic Filters on Microbiological Quality of Drinking Water," *Appl. Environ. Microbiol.* 41:646–651 (1981).
21. Hodges, G. R., J. N. Fink, and N. P. Schlaeter. "Hypersensitivity Pneumonitis Caused by a Contaminated Cool-Mist Vaporizer," *Ann. Int. Med.* 80:501–504 (1974).
22. Metzger, W. J., R. Patterson, J. Fink, R. Semerdjian, and M. Roberts. "Sauna-Taker's Disease. Hypersensitivity Pneumonitis due to Contaminated Water in a Home Sauna," *J. Am. Med. Assoc.* 236:2209–2211 (1976).
23. Jackson, R. A. "Keeping a Century-Old Water System Young," *J. Am Water Works Assoc.* 72:492–495 (1980).

24. Burman, N. P. "Symposium on Consumer Complaints. 4. Taste and Odour due to Stagnation and Local Warming in Long Lengths of Piping," *Water Treatment Exam.* 14:125–131 (1965).
25. Burman, N. P., and J. S. Colbourne. "Effect of Non-Metallic Materials on Water Quality," *J. Inst. Water Eng. Scient.* 33:11–18 (1979).
26. Burman, N. P., and J. S. Colbourne. "Techniques for the Assessment of Growth of Micro-Organisms on Plumbing Materials Used in Contact with Potable Water Supplies," *J. Appl. Bacteriol.* 43:137–144 (1977).
27. Goshko, M. A., W. O. Pipes, and R. R. Christian. "Coliform Occurrence and Chlorine Residual in Small Water Distribution Systems," *J. Am. Water Works Assoc.* 75:371–376 (1983).
28. *Standard Methods for the Examination of Water and Wastewater,* 15th ed. (Washington, DC: American Public Health Association, 1981).
29. Buck, J. D., and P. M. Bubacis. "Membrane Filter Procedure for Enumeration of *Candida albicans* in natural waters," *Appl. Environ. Microbiol.* 35:237–242 (1978).
30. Quereshi, A. A., and B. J. Dutha. "Comparison of Various Brands of Membrane Filters for their Ability to Recover Fungi from Water," *Appl. Environ. Microbiol.* 32:445–451 (1976).
31. Larone, D. H. *Medically Important Fungi: A Guide to Identification* (New York: Harper and Row, 1976).
32. Lodder, J. *The Yeasts: A Taxonomic Study* (Amsterdam: North-Holland Publishing Company, 1970).
33. Raper, K. B., and C. Thom. *A Manual of the Penicillia* (Baltimore: The Williams and Wilkins Company, 1949).
34. Thom, C., and K. B. Raper. *Manual of the Aspergilli* (Baltimore: The Williams and Wilkins Company, 1945).
35. Barnett, H. L. *Illustrated Genera of Fungi Imperfecti* (Minneapolis: Burgess Publishing Company, 1960).

SECTION IV

Chlorination Pathways and By-Products

CHAPTER 8

Pathways for the Production of Organochlorine Compounds in the Chlorination of Humic Materials

Ed W. B. de Leer and Corrie Erkelens, Department of Analytical Chemistry, Delft University of Technology, Delft, The Netherlands

INTRODUCTION

The production of organochlorine compounds in the chlorination of drinking water is well known. The volatile trihalomethanes $CHCl_3$, $CHBrCl_2$, $CHBr_2Cl$, and $CHBr_3$, and the nonvolatile acids dichloroacetic acid (DCA) and trichloroacetic acid (TCA) are the major products that have been detected.[1] It is generally accepted that the reaction between chlorine and aqueous humic material (HM) is responsible for the production of the major portion of these compounds.[2] Humic acids (HA) and fulvic acids (FA) show a high reactivity toward chlorine and constitute 50–90% of the total dissolved organic carbon (DOC) in river and lake waters.[3] Other fractions of the DOC comprise the hydrophilic acids (up to 30%), carbohydrates (ca. 10%), carboxylic acids (ca. 5%), and amino acids (ca. 5%). The reactivity of carbohydrates and carboxylic acids toward chlorine is low, and they are not expected to contribute to the production of organochlorine compounds. However, hydrophilic acids such as citric acid[4] and amino acids[5] will react with chlorine to produce chloroform and other chlorinated compounds, and therefore may contribute to the total organochlorine production.

The study of pathways for the production of organochlorine compounds in aqueous medium has therefore concentrated on the reactions of chlorine with HM and model compounds for HM. Christman et al.[6,7] identified chloroform, chloral, di- and trichloroacetic acid, and 2,2-dichlorobutanedioic acid as the major chlorination products of aqueous HM, which accounted for 53% of the total organic halogen.

A large number of minor products were detected which included several α-chlorinated alkanoic acids and nonchlorinated benzene carboxylic acids. De Leer et al.[8,9] have extended these studies to incorporate chloroform intermediates, chlorinated aromatic acids, and cyano compounds as potential products in drinking water. The most important chlorination products are summarized in Table 1.

Rook[2] proposed resorcinol structures to be the major precursor structure in HM for chloroform production. In accordance with this hypothesis, in the chlorination of terrestrial[8] and aquatic[9] HM, a large series of intermediates were detected that contained a trichloromethyl group and that could be converted into chloroform by further oxidation and/or substitution reactions. A number of these intermediates could be detected also in the chlorination of 3,5-dihydroxybenzoic acid, a partial oxidation product of humic material,[10] confirming the important role of resorcinol structures.

However, the production of chlorinated compounds such as dichloropropanedioic acid and 2,2-dichlorobutanedioic acid or the cyano-substituted acids cannot be explained on the basis of resorcinol structures, and possible production pathways may require protein-type precursors. This chapter describes an investigation into different pathways for the production of organochlorine compounds by using $KMnO_4$ preoxidation of humic acid to destroy resorcinol precursor structures.

EXPERIMENTAL

Most of the experimental details have been published elsewhere and only new experimental procedures are described. The chlorination of humic materials and model compounds, and the quantitative determination of chloroform, DCA, and TCA have been described.[8] The chlorination of amino acids and the determination of the chlorine demand corrected for the production of chlorate have been described.[9] The production of CO_2 as the result of chlorination of model compounds has been described.[11]

$KMnO_4$ Oxidation of Humic Acid

Humic acid 180 mg (terrestrial origin, Liesselse Peel, The Netherlands) was dissolved in 25 mL of 0.1 M NaOH, and diluted with organic free water to 250 mL. After 16 hr, the solution was centrifuged and filtered over a glass fiber filter. The resulting solution was used as a humic acid standard solution for the $KMnO_4$ oxidations.

pH 10.4 Oxidation

A series of oxidations were performed by adding 2, 5, 10, 20, or 30 mL of $KMnO_4$ solution (3.2 g/L) to a mixture of 25 mL of humic acid standard

Table 1. Major Chlorination Products of Humic Materials

Volatile products	$CHCl_3$, CCl_3—CHO, $CHCl_2$—CN
Non-volatile products	
Chlorinated monobasic acids	$CHCl_2$—COOH, CCl_3—COOH,
	CH_3—CCl_2—COOH, CCl_2=CCl—COOH
Chlorinated dibasic acids	HOOC—CCl_2—COOH,
	HOOC—CHCl—CH_2—COOH,
	HOOC—CCl_2—CH_2—COOH,
	HOOC—CCl=CH—COOH,
	HOOC—CCl=CCl—COOH
Chlorinated tribasic acids	HOOC—CCl=C=$(COOH)_2$
Chloroform intermediates	CCl_3—CO—CCl=C=$(COOH)_2$
	CCl_3—CHOH—CCl_2—CHCl—COOH
Cyanoalkanoic acids	NC—CH_2—CH_2—COOH,
	NC—CH_2—CH_2—CH_2—COOH
Alkanoic acids	CH_3—$(CH_2)_n$—COOH n = 6 − 22
Alkanedioic acids	HOOC—$(CH_2)_n$—COOH n = 0 − 2
Benzene carboxylic acids	Phenyl—$(COOH)_n$ n = 1 − 6

solution and 10 mL of $0.25\,M$ pH 10.4 carbonate buffer. When necessary, the pH was corrected by adding $2\,M$ HCl or $2M$ NaOH. After a reaction time of 2 hr, the solutions were acidified to pH 4 with a $4\,M\ H_2SO_4$ solution. Any residual of $KMnO_4$ was reduced with a 5% sodium arsenite solution with amperometric indication (Pt-electrodes, 200 mV, 50 μA). The pH of the resulting solution was increased to precipitate the MnO_2, which was removed by filtration over a glass fiber filter. Carbonate was removed by acidification to pH 3 with concentrated H_2SO_4 and passing nitrogen gas through the solution for 10 min. The pH was adjusted finally to 7.2 with $2\,M$ NaOH, and the volume of the solution was made up to 100 mL with organic free water. After analyzing for the total organic content, the solutions were used for the chlorination experiments at pH 7.2. The $KMnO_4$ oxidations at pH 7.2 were conducted in an analogous manner, except that instead of the carbonate buffer, 4 mL of a $1\,M$ pH 7.2 phosphate buffer was used.

RESULTS AND DISCUSSION

Resorcinol Model Studies

The reaction of chlorine with resorcinol structures in HM may explain the production of chloroform. Resorcinol derivatives (Table 2) gave a chloroform production yield of 72–97% when the ring position between the two hydroxyl groups was unsubstituted. However, substitution of this 2-position with a hydroxyl or a methyl group reduced the chloroform yield to 0–8%. Substitution of the 2-position with a carboxyl group gave no significant reduction in chloroform yield, indicating that the carboxyl group can be

Table 2.   Chlorination of Resorcinol Derivatives. Chlorine Demand and Production of Chloroform and Carbon Dioxide

Compound	Chlorine Demand (M/M)	Production of:	
		$CHCl_3$ (M/M)	CO_2 (M/M)
Resorcinol	7.1	0.78	1.02
2-Methylresorcinol	6.8	0.08	0.99
5-Methylresorcinol	6.8	0.85	—
2,4-Dihydroxybenzoic acid	6.9	0.85	1.87
2,6-Dihydroxybenzoic acid	6.2	0.75	1.87
3,5-Dihydroxybenzoic acid	6.8	0.72	0.64
1,2,3-Trihydroxybenzene	6.5	0.00	2.03
1,3,5-Trihydroxybenzene	9.4	0.97	2.94

expelled through oxidation and substitution reactions.[12] The production of almost 2 moles of carbon dioxide in the chlorination of 1 mole of 2,6-dihydroxybenzoic acid is in agreement with this reaction.

The major chlorination products of resorcinol are $CHCl_3$, CO_2, and chlorobutenedioic acid.[11,13] When the 5-position is substituted, the substituent appears as a substituent in the chlorobutenedioic acid. 5-Methylresorcinol and 3,5-dihydroxybenzoic acid gave, respectively, 2-chloro-3-methylbutenedioic acid and chloroethenetricarboxylic acid as one of the major products. The mechanism for the chlorination of resorcinol was proposed originally by Rook[2] and later extended by Boyce and Hornig.[14] It included a pentachlororesorcinol intermediate (Scheme 1), which after a nucleophilic ring opening reaction was converted into chloroform and several other products, such as 2,4,4-trichloro-2-pentenedioic acid.

$$^*CHCl_3 + HOOC-CCl=CH-CCl_2-COOH \quad (1)$$

However, the production of chlorobutenedioic acid as a major product cannot be explained on the basis of this mechanism and is better explained by a chlorinated p-benzoquinone-type intermediate as proposed by Lin et al.[15] The formation of p-benzoquinones as the result of the chlorination of resorcinol derivatives has also been detected by Onodera et al.[16] and may explain the production of $CHCl_3$, CO_2, and chlorobutenedioic acid by a series of reactions similar to the reactions proposed for the ring opening of the pentachlororesorcinol intermediate as shown in Scheme 2.

$$CHCl_3 + CO_2 + HOOC - CCl = CH - COOH \quad (2)$$

The mechanism based on chlorinated p-benzoquinone intermediates may explain several of the major HM chlorination products listed in Table 1. For example, the production of chloroethenetricarboxylic acid could be explained by this mechanism (Scheme 3). Alternative ring opening mechanisms may lead to DCA and TCA.

$$CHCl_3 + CO_2 + \quad (3)$$

$$(CHCl_2 - COOH ; CCl_3 - COOH)$$

KMnO$_4$ Oxidation Studies

However, the presence of resorcinol-type structures in HM is uncertain because direct instrumental or chemical evidence for these structures is missing. Indirect evidence for these structures comes from the work of Norwood et al.,[17] who detected 3,5-dihydroxybenzoic acid as one of the products in the partial oxidation of HM. Strong oxidants such as KMnO$_4$ will destroy resorcinol structures and therefore should reduce or completely eliminate the production of chloroform from HM.

Addition of 1.5 molar equivalents of KMnO$_4$ to a resorcinol solution indeed destroyed the resorcinol completely, as evidenced by a determination of residual resorcinol with HPLC (Figure 1).

However, complete elimination of the chloroform production required the addition of approximately 6 molar equivalents, and the reduction in chloroform production was in better agreement with the reduction in total phenol content, as found from the nonspecific determination with 4-aminoantipyrine.[18]

The other resorcinol derivatives from Table 2 reacted instantaneously with KMnO$_4$ at pH 10.4 and completely consumed more than 3 molar equivalents. The residual amount of "phenolic structure" as measured with the 4-aminoantipyrine method after the addition of 3 and 8 molar equivalents of KMnO$_4$ is given in Table 3.

Oxidation of terrestrial HA was performed at pH 7.2 and 10.4 with increasing amounts of KMnO$_4$ to ensure complete oxidation of resorcinol

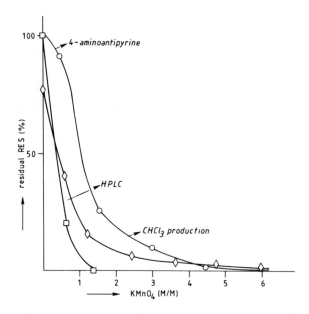

Figure 1. Chloroform production from resorcinol as a function of $KMnO_4$ preoxidation. Residual resorcinol and nonspecific phenols were determined with HPLC and 4-aminoantipyrine, respectively.

structures. $KMnO_4$ was completely consumed up to 2 mg $KMnO_4$/mg C. With larger $KMnO_4$ doses, the residual amount present after a reaction time of 2 hr was reduced by an amperometric titration with sodium arsenite. The oxidative action of KMnO4 was evidenced by a decrease in dissolved organic carbon by approximately 35% at the maximum $KMnO_4$ dose (Figure 2). Partial oxidation of HA to CO_2 may explain this reduction. At the highest $KMnO_4$ dose level the chlorine demand at pH 7.2 (Figure 3) was decreased

Table 3. Residual "Phenolic Content" of Different Resorcinol Derivatives After Oxidation with $KMnO_4$ at pH 10.4

Compound	Residual % Phenolic Content	
	$KMnO_4$ Added (Mol Equiv.)	
	2.7	8.0
2-Methylresorcinol	2.0	ND[a]
5-Methylresorcinol	1.6	ND
2,4-Dihydroxybenzoic acid	27	2.8
2,6-Dihydroxybenzoic acid	6.5	ND
3,5-Dihydroxybenzoic acid	3.4	ND
1,2,3-Trihydroxybenzene	13	ND
1,3,5-Trihydroxybenzene	1.2	ND

[a]ND = not detectable.

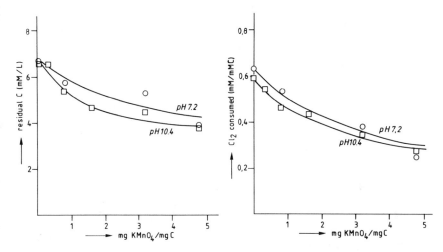

Figures 2 and 3. Reduction in dissolved organic carbon and chlorine demand of HA at pH 7.2 after oxidation with $KMnO_4$ at pH 7.2 and 10.4.

by approximately 50%, and we assume that under these conditions a complete oxidation of resorcinol structures has occurred. No significant differences were observed for the oxidations at the different pH values.

$KMnO_4$ oxidation of the HA decreased the production of chloroform and TCA. At a $KMnO_4$ dose level of 5 mg/mg C both the chloroform and the TCA production decreased to 50% of the original level (Figures 4 and 5).

However, if our assumption of a complete oxidation of resorcinol structures at the 5 mg $KMnO_4$/mg C level is correct, other precursor structures that survive strong oxidation conditions must be present. These structures remain unknown, but amino acids and protein structures are a good candidate because they are not readily oxidized with $KMnO_4$.[19]

The precursor for DCA is rather insensitive to oxidation because we found no significant decrease in its production with increasing $KMnO_4$ doses. A similar result was found by Reckhow and Singer,[20] who detected no decrease in DCA production after ozonation of aquatic HA, while the production of chloroform and TCA decreased significantly.

Resorcinol structures in HM may therefore explain part of the organochlorine production, but other precursor structures are important. Moreover, important chlorination products such as dichloropropanedioic acid, 2,2-dichlorobutanedioic acid, and the cyano-substituted alkanoic acids cannot be explained on the basis of resorcinol precursor structures.

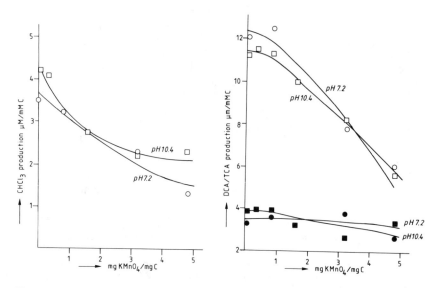

Figures 4 and 5.   Production of chloroform, DCA, and TCA in the chlorination of preoxidized (pH 7.2 or 10.4) humic acid at pH 7.2.

Amino Acid Chlorination Products

We therefore explored several amino acids as precursors for organochlorine compounds. Glutamic acid (Scheme 4) was shown to produce 3-cyanopropanoic acid in high yield when chlorinated at pH 7.2, while lysine gave 4-cyanobutanoic acid together with a dichlorocyanobutanoic acid.[9]

$$H_2N - CH - CH_2 - CH_2 - COOH \xrightarrow{HOCl} N{\equiv}C - CH_2 - CH_2 - COOH + CO_2 \quad (4)$$
$$\overset{|}{\underset{COOH}{}}$$

The chlorination of proline was especially interesting, since it gave 3-cyanopropanoic acid together with dichloropropanedioic acid and 2,2-dichlorobutanedioic acid as the major products. Proline therefore represents the first precursor for some of the chlorination products that cannot be explained on the basis of resorcinol-type compounds (Scheme 5).

$$\begin{array}{c} CH_2 - CH_2 \\ / \qquad \backslash \\ CH_2 \qquad CH - COOH \xrightarrow{HOCl} HOOC - CCl_2 - CH_2 - COOH \quad (5) \\ \backslash_{N}/ \\ \underset{H}{} \end{array}$$
$$+many\ other\ products$$

Aspartic acid, glutamic acid, and uracil also gave small amounts of dichloropropanedioic acid or 2,2-dichlorobutanedioic acid.

The possibility of protein material as an important precursor for organochlorine compounds in the chlorination of humic acid needs to be investigated. Humic materials contain 1–3 weight percent of nitrogen,[3] which is equivalent with 5–15 weight percent of protein material or amino acids. When the reactivity of proteins and amino acids under normal chlorination conditions can be demonstrated, they may explain several of the chlorination products that cannot be understood on the basis of resorcinol structures.

REFERENCES

1. Johnson, J. D., and J. N. Jensen. "THM and TOX Formation: Routes, Rates, and Precursors," *J. Am. Water Works Assoc.* 78(4):156–162 (1986).
2. Rook, J. J. "Chlorination Reactions of Fulvic Acids in Natural Waters," *Environ. Sci. Technol.* 11:478–482 (1977).
3. Thurman, E. M. "Dissolved Organic Compounds in Natural Waters," in *Organic Carcinogens in Drinking Water,* N. M. Ram et al., Eds. (New York: John Wiley and Sons, Inc., 1986), 55–92
4. Larson, R. A., and A. L. Rockwell. "Chloroform and Chlorophenol Production by Decarboxylation of Natural Acids During Aqueous Chlorination," *Environ. Sci. Technol.* 13:325–329 (1979).
5. Trehy, M. L., R. A. Yost, and C. J. Miles. "Chlorination Byproducts of Amino Acids in Natural Waters," *Environ. Sci. Technol.* 20:1117–1122 (1986).
6. Johnson, J. D., R. F. Christman, D. L. Norwood, and D. S. Millington. "Reaction Products of Aquatic Humic Substances with Chlorine," *Environ. Health Pers.* 46:63–71 (1982).
7. Christman, R. F., D. L. Norwood, D. S. Millington, and J. D. Johnson. "Identity and Yields of Major Halogenated Products of Aquatic Fulvic Acid Chlorination," *Environ. Sci. Technol.* 17:625–628 (1983).
8. De Leer, E. W. B., J. S. Sinninghe Damsté, C. Erkelens, and L. de Galan. "Identification of Intermediates Leading to Chloroform and C-4 Diacids in the Chlorination of Humic Acid," *Environ. Sci. Technol.* 19:512–522 (1985).
9. De Leer, E. W. B., T. Baggerman, C. Erkelens, and L. de Galan. "The Production of Cyano Compounds on Chlorination of Humic Acid. A Comparison Between Terrestrial and Aquatic Material," *Sci. Total Environ.* 62:329–334 (1987).
10. Christman, R. F., and M. Ghassemi. "Chemical Nature of Organic Color in Water," *J. Am. Water Works Assoc.* 58:723–741 (1966).
11. De Leer, E. W. B., and C. Erkelens. "Chloroform Production from Model Compounds of Aquatic Humic Material. The Role of Pentachlororesorcinol as an Intermediate," *Sci. Total Environ.* 47:211–216 (1985).
12. Larson, R. A., and A. L. Rockwell. "Gas Chromatographic Identification of Some Chlorinated Aromatic Acids, Chlorophenols, and Their Aromatic Acid Precursors," *J. Chromatog.* 139:186–190 (1977).

13. De Laat, J., N. Merlet, and M. Doré. "Chlorination of Organic Compounds: Chlorine Demand and Reactivity in Relationship to the Trihalomethane Formation," *Water Res.* 16:1437–1450 (1982).
14. Boyce, S. D., and J. F. Hornig. "Reaction Pathways of Trihalomethane Formation from the Halogenation of Dihydroxyaromatic Model Compounds for Humic Acid," *Environ. Sci. Technol.* 17:202–211 (1983).
15. Lin, S., R. J. Liukkonen, R. E. Thorn, J. G. Bastian, M. T. Lukasewycz, and R. M. Carlson. "Increased Chloroform Production from Model Compounds of Aquatic Humus and Mixtures of Chlorine Dioxide/Chlorine," *Environ. Sci. Technol.* 18:932–935 (1984).
16. Onodera, S., M. Tabata, S. Suzuki, and S. Ishikura. "Gas Chromatographic Identification and Determination of Chlorinated Quinones Formed During Chlorination of Dihydric Phenols with Hypochlorite in Dilute Aqueous Solution," *J. Chromatog.* 200:137–144 (1980).
17. Norwood, D. L., R. F. Christman, and P. G. Hatcher. "Structural Characterization of Aquatic Humic Material. 2. Phenolic Content and Its Relationship to Chlorination Mechanism in an Isolated Aquatic Fulvic Acid," *Environ. Sci. Technol.* 21:791–798 (1987).
18. Method 510C, "Direct Photometric Method," in *Standard Methods for the Examination of Water and Wastewater,* 16th ed. (Washington, DC: American Public Health Association, 1985), 560–561.
19. Spicher, R. G., and R. T. Skrinde. "Effect of Potassium Permanganate on Pure Organic Compounds," *J. Am. Water Works Assoc.* 57:472–484 (1965).
20. Reckhow, D. A., and P. C. Singer. "Mechanisms of Organic Halide Formation During Fulvic Acid Chlorination and Implications with Respect to Preozonation," in *Water Chlorination: Chemistry, Environmental Impact and Health Effects, Vol. 5,* R. L. Jolley, R. J. Bull, W. P. Davis, S. Katz, M. H. Roberts, Jr., and V. A. Jacobs, Eds. (Chelsea: Lewis Publishers, Inc., 1985), 807–819.

CHAPTER 9

GC/MS Identification of Mutagens in Aqueous Chlorinated Humic Acid and Drinking Waters Following HPLC Fractionation of Strong Acid Extracts

W. Emile Coleman,* Jean W. Munch, Paul A. Hodakievic, Frederick C. Kopfler, and John R. Meier, U.S. Environmental Protection Agency, Cincinnati, Ohio

Robert P. Streicher and Hans Zimmer, University of Cincinnati, Cincinnati, Ohio

INTRODUCTION

Previous studies conducted at this laboratory on the chlorination of humic acid as a model substrate for dissolved organic carbon in drinking waters[1,2] have revealed the formation of numerous chlorination by-products, many of which have shown mutagenic activity in the Ames test.[3,4] Most of the mutagenic activity of chlorinated humic acid (CHA) solutions could be traced to the strong acid fraction.[5] Diethyl ether extracts of these strong acid fractions of CHA were further fractionated using high-pressure liquid chromatography (HPLC) separation. Drinking waters of three U.S. cities were concentrated by Rohm and Haas Amberlite XAD adsorption/acetone elution followed by strong acid extraction and HPLC separation. Meier et al.[5,6] reported that the mutagenic activities of CHA and of the three drinking water samples after HPLC separation were each concentrated in the same two subfractions. GC/MS analyses using fused silica capillary GC columns were performed on the mutagenic HPLC subfractions of CHA and the three drinking water samples. This chapter reports the tentative identification of compounds and discusses the similarity of mutagenic disinfection by-products in CHA and drinking waters.

---

*To whom inquiries should be addressed.

EXPERIMENTAL

Chlorination of solutions of humic acid (1 g/L TOC) was done as previously described by Coleman et al.[3] Isolation of mutagens from these solutions has been described by Meier et al.[5] Organic substances were collected from three chlorinated drinking waters, which had been immediately acidified to pH 2, by passing each through columns of Amberlite XAD8 and XAD2 resins, and then eluting the columns with acetone. These acetone eluates were evaporated, dissolved in diethyl ether, and the strong acids were isolated using sodium bicarbonate extraction. The strong acid extracts were then fractionated further into 1-mL aliquots using HPLC on a C_{18} column using a methanol in water gradient. Details of the drinking water concentration and separation into mutagenic fractions are described by Meier et al.[6] The importance of sample pH on recovery of mutagenicity from drinking water by Amberlite XAD resins is reported by Ringhand et al.[7] Table 1 shows the characteristics of the three drinking water samples.[8]

HPLC fractions containing mutagenic activity, as identified by the *S. typhimurium* reversion assay, were methylated[6] and analyzed by GC/MS via the following conditions:

Gas Chromatograph: Carlo Erba 4160

Column: 15 m × 0.25 mm i.d. SE-54

Injection: 2 μL splitless; splitless period 30 seconds

Injector Temperature: 250°C

Oven Temperature Program: 30°C for 2 minutes; then program at 8°C/minute to 250°C; hold 20 minutes

Mass Spectrometer: Finnigan 3300, Electron Impact

Scan Range/Rate: 14–750 amu in 1 second

3-Chloro-4-(dichloromethyl)-5-hydroxy-2(5H)-furanone (MX) and 5,5,5-trichloro-4-oxopentanoic acid were quantified as their methyl derivatives by comparison of their peak areas to a standard curve derived from peak areas of known amounts of each authentic standard. Single ion chromatograms of m/z 147 and m/z 115 were used, respectively.

MX was synthesized by the method of Padmapriya et al.,[9] except that aluminum chloride was substituted for ferric chloride in the chlorination reaction. 5,5,5-Trichloro-4-oxopentanoic acid and 5-hydroxy-5-trichloromethyl-2-furanone were prepared as described by Winston et al.[10]

Table 1. Water Characteristics of Three Drinking Water Utilities

Utility Code	Water Source	Typical Concentration Organic Carbon (mg/L)	Free Residual Cl (mg/L)	Drinking Water pH
City 1	Surface	20–25 (R) 5–10 (F)	3.2	8.6
City 2	Shallow Ground	23 (R) 9 (F)	0.2	7.2
City 3	Surface	1–3 (R) <2–2.5 (F)	1.4	8.5

Source: Stevens et al.[8]
(R) = raw water; (F) = finished water.

RESULTS AND DISCUSSION

Diethyl ether extracts of the strong acid fraction of CHA were subjected to HPLC separation followed by a mutagenesis assay on the subfractions. Examination of the data from the collection of HPLC subfractions for additional mutagenicity testing showed that the activity was mainly concentrated in two subfractions, 16 and 19, as shown in Figure 1. The solid line trace indicates the mutagenic potency of each subfraction. Table 2 shows the results of GC/MS analysis of combined HPLC fractions 15 and 16. When the total ion chromatogram from a mixture of methylated HPLC fractions 15 and 16 (Figure 2) are compared with the list of compounds in Table 2, it can be seen that the major components are C_4-C_6 aliphatic chlorinated carboxylic, dicarboxylic, and keto acids. Due to the unavailability of authentic commercial or synthesized standards, except where indicated, these tentative identifications are based on our interpretation of the mass spectra. De Leer et al.[11,12] and Lindstrom and Oesterberg[13] have reported very similar by-products from the chlorination of humic materials and Kraft pulp spent bleach liquors, respectively. Among the chemicals identified in fractions 15 and 16 was the potent bacterial mutagen MX,[9,14-16] which was present at a concentration of 28 µg/L of CHA.[6] The value 28 µg/L is an average of several analyses, but it is not corrected for recovery and it is the result of a single chlorination experiment. The majority of MX occurred in fraction 16, although it was identified at low levels in fraction 15. Fractions 15 and 16 were combined to facilitate the quantitation of MX in the total strong acid fraction.

The compounds identified by GC/MS analysis of fraction 19 are shown in Table 3. Most of these compounds were C_5-C_6 chlorinated keto acids containing 3 to 5 chlorines. Five compounds in fraction 19 were also detected in the analysis of the pool of fractions 15 and 16; their concentrations (based

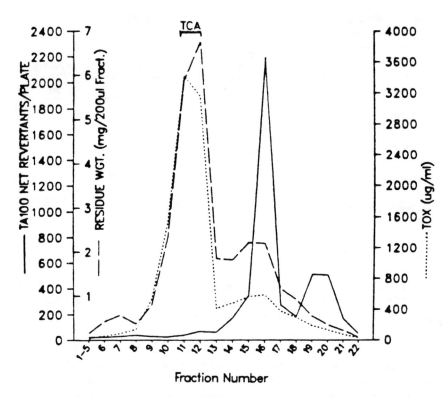

Figure 1. Semi-preparative HPLC analysis of the strong acid fraction of chlorinated humic acid. TCA is trichloroacetic acid, which comprises most of the TOX and the residue weight of the ether extract.[5]

on area counts) were $8\times$ to $1000\times$ higher in combined fraction 15 and 16. The reason for this carryover into fraction 19 is unknown; but it is possible that resolution of the semi-preparative column was poor or the column was overloaded. The average elemental chlorine content, based on the molecular weights of the tentatively identified compounds in mutagenic HPLC fractions 15–16 and 19, is about 50%.

Ether extracts of strong acid fractions from the three United States drinking water samples were similarly subjected to the same HPLC separation followed by mutagenesis testing of the HPLC subfractions. As with the CHA fractions, the mutagenic activity of each drinking water sample was concentrated in fractions 16 and 19.[6] Likewise, MX was identified in fractions 15 and 16 of each of these drinking water samples. In addition, with the CHA concentrates, the majority of MX was in fraction 16. The concentrations of MX for City 1, City 2, and City 3 were 33, 18, and 2 ng/L, respectively. These values were not corrected for recovery efficiency, so they indicate a minimum concentration of MX present in each water sample. Recent recov-

Table 2. Tentative Identification of Compounds in Methylated HPLC Fractions 15 and 16 of the Strong Acid Fraction of Chlorinated Humic Acid

GC Peak No.[a]	Compound Name	Proposed Structure	Mol. Wt.[b]	EI Mass Spectra[c] m/z (% Relative Abundance)
1	1,1,3,3-Tetrachloropropanone[d,e,f,g,h]	$CHCl_2$—CO—$CHCl_2$	194	83(100), 85(81), 76(28), 111(25) 48 (23), 113(17), 87(15), 47(13)
2	1,1,1,3-Tetrachloropropane[d,e,f,g]	CCl_3—CO—CH_2Cl	194	77(100), 79(30), 49(27), 117(26) 119 (25), 82(24), 47(20), 131(19)
3	Trichloropropanoic acid	$C_2H_2Cl_3$—COOH	176	96(100), 59(63), 98(59), 61(24) 133 (23), 89(22), 15(19), 97(16)
4	2,3,3-Trichloropropenoic acid	C_2Cl_3—COOH	174	157(100), 87(92), 131(73), 159(72) 94 (50), 59(49), 153(47), 89(38)
5	Pentachloropropanone[d,e]	CCl_3—CO—$CHCl_2$	228	83(100), 117(84), 119(84), 85(63) 111 (32), 82(25), 47(23), 113(20)
6	Benzoic acid[d,f,g,h]	⬡—COOH	122	105(100), 77(64), 136(28), 51(12) 106 (9), 50 (7), 92 (4), 74 (3)
7	2-Chlorobutenedioic acid[d]	HOOCCICl=CH—COOH	150	147(100), 149(38), 59(18), 53 (9) 69 (8), 15 (8), 60 (6), 29 (5)
8	2,2- or 2,3- or 3,3-Dichloro-4-oxopentanoic acid[f,g,h]	CH_3CO—$C_2H_2Cl_2$—COOH	184	43(100), 131(28), 59(15), 15(14) 61 (12), 133(12), 89 (8), 167 (8)
9	2,3-Dichloro-4-oxopentenoic acid	CH_3CO—ClC=CCl—COOH	182	165(100), 117(80), 43(68), 87(64) 167 (62), 161(61), 181(57), 89(40)
10	Trichloro-4-oxo-pentenoic acid	$CHCl_2$—CO—CH_2Cl—COOH	216	83(100), 85(67), 76(27), 48(26) 147 (21), 105(16), 87(15), 59(12)
11	3,3,3-Trichloro-2-hydroxy-2-methylpropanoic acid	HOOC—C(OH)(CH_3)—CCl_3	206	43(100), 125(70), 161(50), 163(48) 127 (47), 103(44), 126(40), 97(30)

Table 2. Continued

GC Peak No.[a]	Compound Name	Proposed Structure	Mol. Wt.[b]	EI Mass Spectra[c] m/z (% Relative Abundance)
12	2-Methylpentanedioic acid	HOOC—CH(CH$_3$)—CH$_2$—CH$_2$—COOH	146	114(100), 115(64), 59(38), 55(36), 73 (33), 143(30), 99(28), 88(27)
13	Isomer of #7[d]		150	147(100), 149(37), 59(14), 53 (9), 15 (8), 69 (7), 119 (7), 87 (6)
14	2,2- or 2,3-Dichlorobutanedioic acid	HOOC—CH$_2$Cl$_2$—COOH	186	59(100), 135(72), 155(43), 61(33), 15 (32), 157(27), 183(24), 137(20)
15	Trichlorooxobutanoic acid	C$_2$H$_2$Cl$_3$CO—COOH	204	159(100), 161(97), 95(36), 163(35), 97 (20), 89(13), 183(10), 107 (8)
16	Hexanedioic acid	HOOC(CH$_2$)$_4$COOH	146	114(100), 101(84), 111(82), 59(76), 143 (72), 115(69), 55(58), 15(36)
17	5-hydroxy-5-trichloromethyl-2-furanone[d,e,g]		216	113(100), 85(17), 82(11), 59 (9), 53 (5), 117 (4), 114 (4), 54 (4)
18	Trichloro-4-oxo-pentenoic acid	CHCl$_2$—CO—C$_2$HCl—COOH	216	147(100), 83(71), 85(48), 149(37), 59 (10), 48 (9), 87 (8), 53 (7)
19	5,5,5-Trichloro-4-oxopentanoic acid[d,e,f,g,h]	CCl$_3$CO—CH$_2$CH$_2$—COOH	218	115(100), 55(59), 59(48), 87(27), 82 (15), 28(13), 15(13), 109(11)
20	2,5,5- or 3,5,5-Trichloro-4-oxopentenoic acid[f,g,h]	CHCl$_2$—CO—C$_2$HCl—COOH	216	147(100), 149(37), 59(13), 83(11), 85 (8), 87 (8), 76 (6), 60 (6)
21	3,5,5- or 2,5,5-Trichloro-4-oxopentenoic acid[f,g,h]	CHCl$_2$—CO—C$_2$HCl—COOH	216	147(100), 149(36), 83 (5), 143 (5), 59 (5), 145 (4), 60 (4), 69 (4)
22	Isomer of #17		216	113(100), 85(13), 82 (9), 89 (7), 143 (6), 59 (6), 145 (6), 114 (5)
23	Trichloro-4-oxo-methylbutenoic acid	HOOC—CCl=C(CHCl$_2$)—CHO	216	170(100), 172(99), 142(74), 107(52), 144 (49), 109(34), 174(25), 173(21)

Table 2. Continued

GC Peak No.[a]	Compound Name	Proposed Structure	Mol. Wt.[b]	EI Mass Spectra[c] m/z (% Relative Abundance)			
24	Unknown			139(100), 141(66), 174(26), 176(25)	129(27), 73(15), 89(14), 15(14)		
25	3-Chloro-4-(dichloromethyl)-5-hydroxy-2(5H)-furanone (MX)[d,e,f,g,h]	(furanone ring; $HCCl_2$, Cl)	216	147(100), 107(61), 149(33), 137(29)	109(27), 201(26), 29(24), 72(21)		
26	2-Chloro-3-(dichloromethyl)-4-oxopentenoic acid	$CH_3COC(CHCl_2)=CCl-COOH$	230	133(100), 135(28), 43(15), 115(13)	101(9), 69(8), 117(8), 61(6)		
27	Unknown			85(100), 43(78), 117(60), 151(56)	153(37), 183(28), 185(20), 71(17)		
28	2,5,5,5- or 3,5,5,5-Tetrachloro-4-oxopentenoic acid	CCl_3-CO-C_2OOH	250	147(100), 119(39), 149(37), 59(34)	69(19), 121(15), 82(14), 88(12)		
29	Isomer of #28	$CCl_3-CO-C_2HCl-COOH$	250	147(100), 149(34), 133(23), 59(14)	75(13), 87(9), 119(9), 53(8)		
30	Tetrachloro-hydroxypentanoic acid	$HOOC-C_2H_2Cl-CH(OH)-CCl_3$	254	113(100), 151(99), 75(70), 115(68)	117(55), 119(49), 92(40), 153(38)		
31	Pentachloro-hydroxypentanoic acid	$HOOC-C_2H_2Cl_2-CH(OH)-CCl_3$	288	113(100), 141(76), 117(66), 115(61)	143(61), 119(60), 121(22), 59(21)		
32	Isomer of #26	$HOOC-C_3H_3Cl-CO-CHCl_2$	230	133(100), 135(28), 181(15), 107(11)	147(9), 59(9), 159(9), 183(8)		
33	Trichlorohydroxypentanedioic acid	$HOOC-CHCl-CH(OH)-CCl_2-COOH$	250	101(100), 187(60), 189(54), 152(52)	129(40), 154(37), 59(31), 110(28)		
34	Pentachlorooxopentenoic acid	$HOOC-CO-CCl=CCl-CCl_3$	284	141(100), 143(98), 59(82), 106(61)	213(61), 178(47), 211(42), 215(37)		
35	Isomer of #30	$HOOC-C_2H_3Cl-CH(OH)-CCl_3$	254	113(100), 137(72), 151(70), 117(66)	119(63), 115(62), 59(50), 75(42)		
36	Tetrachlorohydroxy-oxohexanoic acid	$HOOC-CHCl-CH(OH)-CH_2-CO-CCl_3$	282	137(100), 159(44), 161(42), 139(29)	117(29), 119(28), 59(31), 71(24)		

Table 2. Continued

GC Peak No.[a]	Compound Name	Proposed Structure	Mol. Wt.[b]	EI Mass Spectra[c] m/z (% Relative Abundance)
37	Nonanedioic acid	HOOC—(CH$_2$)$_7$—COOH	188	152(100), 83(76), 111(56), 124(44) 125 (42), 74(40), 59(38), 185(37)
38	Isomer of #36		282	137(100), 139(41), 105(32), 93(28) 59 (26), 61(25), 117(24), 82(23)
39	Dichlorooctenedioic acid (unmethylated)	ClC=CCl[(CH$_2$)$_2$COOH]$_2$	208	178(100), 143(69), 115(68), 180(58) 150 (30), 145(25), 117(22), 222(20)
40	Pentachlorohydroxybutanoic acid	HOOC—CCl$_2$CH(OH)CCl$_3$	274	127(100), 59(81), 171(78), 129(63) 117 (60), 119(56), 173(51), 112(48)
41	Tetrachlorohydroxypentanedioic acid	HOOC—CCl$_2$—CCl(OH)—CHCl—COOH	284	135(100), 59(49), 183(37), 107(34) 15 (34), 108(34), 137(33), 101(32)
42	Dichloromethyl isomer of #41	HOOC—CCl$_2$—COH(CHCl$_2$)—COOH	284	142(100), 59(55), 144(55), 135(48) 159 (40), 141(38), 15(35), 161(35)
43	2-Carboxy-3,5,5,5-tetrachloro-4-oxopentenoic acid	(HOOC)$_2$—C=CCl—CO—CCl$_3$	294	205(100), 207(34), 87(18), 59(15) 109 (12), 15 (9), 133 (8), 111 (8)
44	3-(Dichloromethyl)-2,5,5,5-tetrachloro-4-oxopentenoic acid (unmethylated)	HOOC—CCl=C(CHCl$_2$)—CO—CCl$_3$	332	179(100), 154(74), 181(72), 119(67) 156 (63), 135(35), 107(33), 121(33)

aCorresponds to peak numbers in Figure 2.
bMolecular weight of underivatized compounds.
cMass spectra of acids are shown as methyl derivatives.
dConfirmed with authentic standards.
eTested positive in Ames Test.
fFound in City-1 drinking water.
gFound in City-2 drinking water.
hFound in City-3 drinking water.

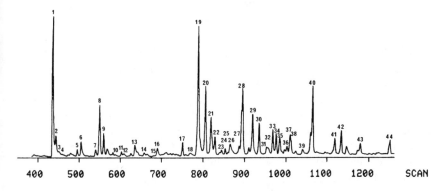

Figure 2. Total ion chromatogram of combined mutagenic HPLC fraction 15 and 16 (methylated) from the strong acid fraction of chlorinated humic acid. The numbers above the GC peaks refer to peak number in Table 2.

ery studies in this laboratory indicate that the actual concentrations in drinking water could be much higher than the amounts reported here. The highest MX concentration was found in the water sample having the highest total organic carbon (TOC) content (see Table 1). The TOC of the model humic acid solution was 1 g/L, whereas the TOC of the source waters of the three drinking water samples ranged from about 3 to 25 mg/L. In addition to the compounds in Table 2, the combined fraction 15 and 16 of the drinking water samples also contained chloral hydrate and benzeneacetic acid.

Meier et al.[6] reported a specific mutagenicity of 13,050 revertants/nmole or 60 revertants/ng in strain TA100 for MX, ranking it among the most potent bacterial mutagens tested.[17] They also reported that MX accounts for

Table 3. Tentative Identification of Compounds in Methylated HPLC Fraction 19 of the Strong Acid Fraction of Chlorinated Humic Acid

Pentachloropropanone
Chlorobenzoic acid
5,5,5-Trichloro-4-oxopentanoic acid[a]
2,5,5-Trichloro-4-oxopentenoic acid[a]
3,5,5-Trichloro-4-oxopentenoic acid[a]
2,3,5,5-Tetrachloro-4-oxopentenoic acid
2,5,5,5-Tetrachloro-4-oxopentenoic acid[a]
3,5,5,5-Tetrachloro-4-oxopentenoic acid[a]
Pentachloropentenoic acid
Tetrachlorooxohexenoic acid
Tetrachlorooxohexanoic acid
Pentachlorooxohexanoic acid
Trichlorocyclohexadienone carboxylic acid
3-Chloro-2-(dichloromethyl)-butenedioic acid

[a]These compounds were also detected in combined HPLC fraction 15 and 16. Their concentrations (based on area counts) were much higher in fraction 15 and 16: 1000×, 60×, 8×, 10×, and 240×, respectively, from top to bottom.

15–34% and 41% of the mutagenicity of XAD concentrates of drinking water and CHA, respectively.

Figure 3 shows GC/MS profiles of methylated fractions 15 and 16 (combined) from a chlorinated drinking water sample (City 1) and a CHA sample. Two compounds in addition to MX that have been identified in CHA solutions and at least one drinking water sample, confirmed with authentic standards, and tested for mutagenicity (unreported data) are: 5-hydroxy-5-trichloromethyl-2-furanone and 5,5,5-trichloro-4-oxopentanoic acid, represented by peaks 17 and 19, respectively, while peak 25 is MX. The mass spectra of these peaks are shown in Figures 4–6. The strain TA100 mutagenic potencies of 5-hydroxy-5-trichloromethyl-2-furanone and 5,5,5-trichloro-4-oxopentanoic acid were about 0.1% and 0.5% that of MX, respectively.[18] The largest acid peak appearing in the chromatogram of combined fraction 15 and 16 of CHA was that of 5,5,5-trichloro-4-oxopentanoic

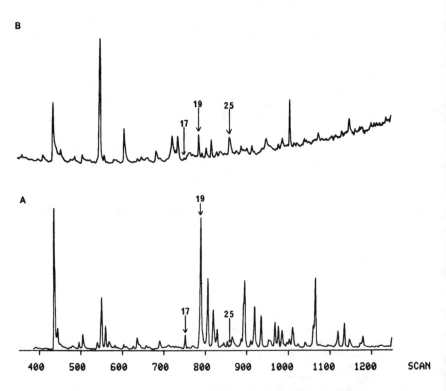

Figure 3. Total ion chromatograms of combined mutagenic HPLC fractions 15 and 16 (methylated) from A, the strong acid fraction of chlorinated humic acid, and B, from the strong acid fraction of City-1 drinking water. Chromatogram A represents 150X concentration of CHA, and chromatogram B, 1,000,000X concentration of City-1 drinking water. Refer to Table 2 for identifications of peaks 17, 19, and 25.

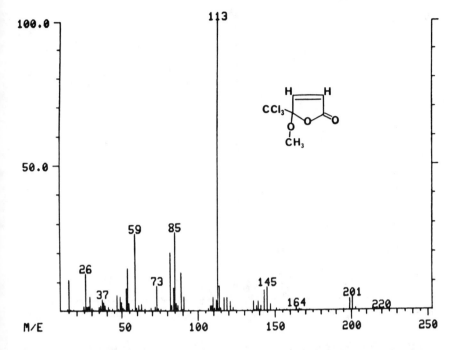

Figure 4. EI mass spectra of GC peak #17, 5-hydroxy-5-trichloromethyl-2-furanone, methyl ether, MW 230.

acid. Its concentration in the CHA solution was 1200 μg/L. Even though 5,5,5-trichloro-4-oxopentanoic acid occurs at a concentration 43 times greater than does MX, because its mutagenic potency is 200 times less than MX, its contribution to the total CHA mutagenicity is about 9% compared to 41% for MX. The mutagenic material in these fractions is direct acting, since it requires no mammalian liver enzyme activation in the bacterial mutagenesis assay.[6]

As more standards are synthesized for the compounds tentatively identified in the mutagenic HPLC fractions, it is likely that a greater percentage of the mutagenic activity of CHA and of drinking water can be accounted for. The presence of mutagenic chlorinated compounds in several drinking waters and in CHA shows that CHA is a good model for predicting disinfection by-products in drinking waters.

CONCLUSIONS

Isolation, fractionation, and derivatization techniques in conjunction with high resolution capillary column GC/MS analysis have proven to be valuable

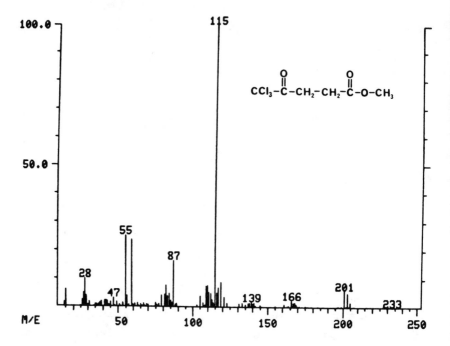

Figure 5. EI mass spectra of GC peak #19, 5,5,5-trichloro-4-oxopentanoic acid, methyl ester, MW 232.

tools in characterizing mutagenic activity in CHA solutions and in drinking waters. These chemical techniques, combined with the mutagenesis bioassay, have provided invaluable information for evaluating the potential health risks posed by the presence of mutagenic compounds in drinking waters, many of which are disinfection by-products. The characterization of additional mutagenic activity in test samples seems to be limited by problems in the analytical methodologies. In spite of improvements in isolation, concentration, and fractionation techniques, it will become necessary in future work to use more advanced mass spectroscopy (MS) techniques such as LC/MS and MS/MS with soft ionization in order to analyze these highly polar and labile constituents directly and quantitatively, thus avoiding derivatization and high-temperature GC. This could include the analysis for macromolecular adducts or other conjugates formed both in vivo and in vitro and with disinfection by-products. The identification of disinfection by-products that are important contributors to the mutagenic activity of drinking water should provide a way of ranking chemicals for further testing in the more expensive animal carcinogenesis bioassays that are currently used to develop regulatory standards for carcinogens. Finally, the results of this study indicate that source waters containing significant humic materials produce, when chlorinated under drinking water treatment conditions, many of the same types of

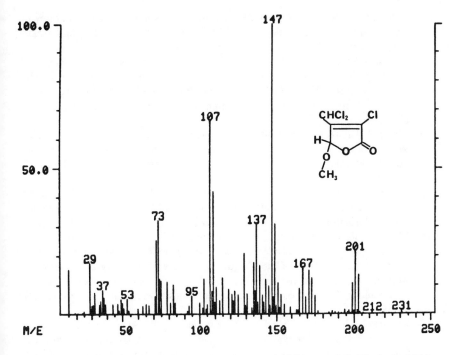

Figure 6. EI mass spectra of GC peak #25, 3-chloro-4-(dichloromethyl)-5-hydroxy-2(5H)-furanone, methyl ether, MW 230.

mutagenic compounds that are found in CHA solutions. These data indicate that additional health effects and treatment studies are necessary in order to determine the quantitative and qualitative impact of both MX-type compounds and chlorinated strong acids on human health.

ACKNOWLEDGMENTS

We thank La Verne Clayton, Linda Ransick, and Pat Underwood for clerical assistance.

This document has been reviewed in accordance with U.S. Environmental Protection Agency policy and approved for publication. Mention of trade names or commercial products does not constitute endorsement or recommendation for use.

REFERENCES

1. Bull, R. J., M. Robinson, J. R. Meier, and J. Stober. "Use of Biological Assay Systems to Assess the Relative Carcinogenic Hazards of Disinfection By-Products," *Environ. Health Pers.* 46:215–227 (1982).

2. Meier, J. R., R. D. Lingg, and R. J. Bull. "Formation of Mutagens Following Chlorination of Humic Acid. A Model for Mutagen Formation During Drinking Water Treatment," *Mutation Res.* 118:25–41 (1983).

3. Coleman, W. E., J. W. Munch, W. H. Kaylor, R. P. Streicher, H. P. Ringhand, and J. R. Meier. "Gas Chromatography/Mass Spectroscopy Analysis of Mutagenic Extracts of Aqueous Chlorinated Humic Acid. A Comparison of the Byproducts to Drinking Water Contaminants," *Environ. Sci. Technol.* 18:674–681 (1984).

4. Meier, J. R., H. P. Ringhand, W. E. Coleman, J. W. Munch, R. P. Streicher, W. H. Kaylor, and K. M. Schenck. "Identification of Mutagenic Compounds Formed During Chlorination of Humic Acid," *Mutation Res.* 157:111–122 (1985).

5. Meier, J. R., H. P. Ringhand, W. E. Coleman, K. M. Schenck, J. W. Munch, R. P. Streicher, W. H. Kaylor, and F. C. Kopfler. "Mutagenic By-Products from Chlorination of Humic Acid," *Environ. Health Pers.* 69:101–107 (1986).

6. Meier, J. R., R. B. Knohl, W. E. Coleman, H. P. Ringhand, J. W. Munch, W. H. Kaylor, R. P. Streicher, and F. C. Kopfler. "Studies on the Potent Bacterial Mutagen, 3-Chloro-4-(dichloromethyl)-5-hydroxy-2(5H)-furanone: Aqueous Stability, XAD Recovery and Analytical Determination in Drinking Water and in Chlorinated Humic Acid Solutions," *Mutation Res.* 189:363–373 (1987).

7. Ringhand, H. P., J. R. Meier, F. C. Kopfler, K. M. Schenck, W. H. Kaylor, and D. E. Mitchell. "Importance of Sample pH on the Recovery of Mutagenicity from Drinking Water by XAD Resins," *Environ. Sci. Technol.* 21:382–387 (1987).

8. Stevens, A. A., L. A. Moore, C. J. Slocum, B. L. Smith, D. R. Seeger, and J. C. Ireland. "By-Products of Chlorination at Ten Operating Utilities," paper presented at the Sixth Conference on Water Chlorination: Environmental Impact and Health Effects, Oak Ridge, Tennessee, May 3–8, 1987.

9. Padmapriya, A. A., G. Just, and N. G. Lewis. "Synthesis of 3-Chloro-4-(dichloromethyl)-5-hydroxy-2(5H)-furanone, a Potent Mutagen," *Can. J. Chem.* 63:828–832 (1985).

10. Winston, A., J. P. M. Bederka, W. G. Isner, P. C. Juliano, and J. C. Sharp. "Trichloromethylation of Anhydrides. Ring-Chain Tautomerism," *J. Org. Chem.* 30:2784–2787 (1965).

11. De Leer, E. W. B., J. S. S. Damste, C. Erkelens, and L. de Galan. "Identification of Intermediates Leading to Chloroform and C-4 Diacids in the Chlorination of Humic Acid," *Environ. Sci. Technol.* 19:512–522 (1985).

12. De Leer, E. W. B., J. S. S. Damste, and L. de Galan. "Formation of Aryl-Chlorinated Aromatic Acids and Precursors for Chloroform in Chlorination of Humic Acid," in *Water Chlorination: Chemistry, Environmental Impact and Health Effects, Vol. 5,* R. L. Jolley, R. J. Bull, W. P. Davis, S. Katz, M. H. Roberts, Jr., and V. A. Jacobs, Eds. (Chelsea, MI: Lewis Publishers, Inc., 1985), 843.

13. Lindstrom, K., and F. Oesterberg. "Chlorinated Carboxylic Acids in Softwood Kraft Pulp Spent Bleach Liquors," *Environ. Sci. Technol.* 20:133–138 (1986).

14. Holmbom, B. R., R. H. Voss, R. D. Mortimer, and A. Wong. "Isolation and Identification of an Ames-Mutagenic Compound Present in Kraft Chlorination Effluents," *TAPPI* 64:172–174 (1981).
15. Holmbom, B. R., R. H. Voss, R. D. Mortimer, and A. Wong. "Fractionation, Isolation, and Characterization of Ames Mutagenic Compounds in Chlorination Effluents," *Environ. Sci. Technol.* 18:333–338 (1984).
16. Hemming, J., B. Holmbom, M. Reunanen, and L. Kronberg. "Determination of the Strong Mutagen 3-Chloro-4-(dichloromethyl)-5-hydroxy-2(5H)-furanone in Chlorinated Drinking and Humic Waters," *Chemosphere* 15:549–556 (1986).
17. McCann, J., L. Horn, and J. Kaldor. "An Evaluation of *Salmonella* (Ames) Test Data in the Published Literature: Application of Statistical Procedures and Analysis of Mutagenic Potency," *Mutation Res.* 134:1–47 (1984).
18. Streicher, R. P., H. Zimmer, J. R. Meier, R. B. Knohl, F. C. Kopfler, W. E. Coleman, J. W. Munch, and K. M. Schenck. "Structure Activity Relationships in Chlorinated α,β-Unsaturated Carbonyl Compounds," poster presented before the Division of Environmental Chemistry at the 193rd National ACS Meeting, Denver, Colorado, April 5–10, 1987.

CHAPTER 10

Some Lipophilic Compounds Formed in the Chlorination of Pulp Lignin and Humic Acids

A. Bruce McKague and Knut P. Kringstad, Swedish Forest Products Research Laboratory, Stockholm, Sweden

INTRODUCTION

Lipophilic compounds have high solubility in fat and low solubility in water. Thus, in the presence of both fat and water matrices, they will partition preferentially into the fat matrix. The logarithm of the partition coefficient, log P, in the two-phase system n-octanol/water is a measure of the degree of lipophilicity and is the parameter normally used to characterize this property of organic compounds.[1,2] Chlorophenols, common industrial pollutants, have log P values in the range 3–5,[1] and are regarded as moderately lipophilic compounds with a propensity for bioaccumulation. Compounds such as DDT and PCB have higher log P values and represent the upper extreme in the range of log P values. A number of chlorinated lipophilic compounds have entered our environment and have been found to persist for many years and to accumulate in aquatic organisms and sediments.

During the conventional bleaching of chemical pulps, between 45 and 90 kg of organic material/ton of pulp are dissolved in the bleaching liquors.[3] Much of the dissolved material is chlorinated and, in the case of softwood kraft pulp, about 5 kg of organically bound chlorine/ton of pulp is produced.[3] It is known that bleaching liquors exert weak acute toxic and genotoxic effects (Ames test), and several compounds that are responsible for such effects have been identified. Although most of the toxic compounds are chemically rather unstable and biodegradable, chlorinated phenols, catechols, and guaiacols originating from pulp bleaching have been shown to accumulate in fish living in the vicinity of pulp mill outlets.[4,5] Dichloromethyl methyl sulfone, a compound often present in spent bleach liquors,[6] has been found in fish and mussels.[7]

In Sweden, a number of steps have been taken to reduce the discharge of chlorinated organics from the pulp industry. Oxygen prebleaching of softwood kraft pulp and increased use of chlorine dioxide have led to a substantial reduction in chlorine consumption.[8] Also, as of 1990, mills will be required to discharge effluents that contain not more than 1.5–2.0 kg TOCl per ton bleached pulp.

The above steps are being taken to reduce the wholesale introduction of chlorinated organics into the environment. In addition, a research program financed by the Swedish pulp industry is currently under way at the Swedish Pulp and Paper Research Institute (STFI) to characterize lipophilic compounds resulting from the bleaching of pulp by present and anticipated future processes. Because of the similarity in the nature of pulp lignin and humic acid and the possibility traces of chlorinated lipophilic compounds may be present in drinking water, work is also being done on humic acid. The work has resulted in the identification of some new chlorinated furanones and has revealed some previously unrecognized degradation pathways for lignin and humic acid.

RESULTS AND DISCUSSION

Spent Chlorination Liquors

The scheme developed for concentration and identification of lipophilic compounds in spent bleach liquors is shown in Figure 1. The liquor was first extracted with hexane to separate nonpolar material from the large amount of polar material present. Lipophilic material was then concentrated by preparative reversed-phase thin layer chromatography (RPTLC). It has recently been shown that partition coefficients derived from RPTLC (R_m values) correlate well with log P values,[9,10] and this method is readily adapted to small-scale preparative work. In our system, plates were developed with acetone:water, 80:20, and the material recovered which had an Rf ≤ benzophenone, a reference compound with log P 3.18.[1] The choice of benzophenone as the lower cutoff was based on recommendations made by the Organization for Economic Co-operation and Development (OECD) for studies of bioaccumulation in living organisms.[2] A similar approach using reversed-phase high-performance liquid chromatography has previously been used for this purpose.[11]

Lipophilic material was then analyzed by gas chromatography with electron capture (GC/ECD) and mass spectrometry (GC/MS) detection using capillary DB1 and DB5 columns. A Finnigan TSQ-46C system was used for GC/MS work. Further fractionation by adsorption chromatography on silica gel was considered optional depending on the complexity of the lipophilic

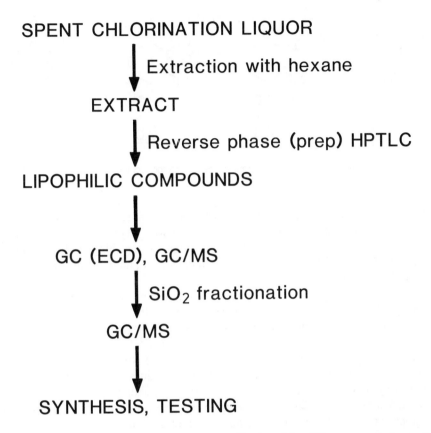

Figure 1.   Scheme for concentration and identification of lipophilic compounds in spent chlorination liquor.

material. Finally, synthesis was employed to aid identification and to provide material for biological and chemical tests.

Initially, spent bleach liquor was prepared in the laboratory from softwood kraft pulp having kappa numbers in the range 30–40.* Pulp was bleached at 3.5% pulp consistency at room temperature with 6.5–7.5% chlorine charge on pulp for 1 hour, then filtered to give the liquor. After extraction with hexane and preparative RPTLC, this gave a lipophilic concentrate containing at least 20 chlorinated compounds, including tetrachlorocyclopentene-1,3-dione *1* and the three furanones *2a, 2b,* and *3* (Figure 2).[12,13] Estimated quantities produced by laboratory bleaching of the pulp ranged from less than 1 g/metric ton pulp for the furanones *2a, 2b,* and *3* to 72 g/ton for

*Kappa number is a measure of lignin remaining in unbleached pulps, characterizing the degree of pulping and the amount of bleach required. It is determined by measuring the consumption of permanganate in treating the unbleached pulp under standardized conditions.

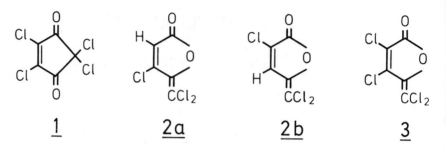

Figure 2.   Compounds identified in lipophilic fraction of spent chlorination liquor.

tetrachlorocyclopentene-1,3-dione. Subsequent analysis of spent chlorination liquor from a Swedish pulp mill that employs oxygen prebleaching to reduce the kappa number to below 20 prior to bleaching with chlorine showed that significantly smaller, but still detectable, amounts of these compounds were present.

The furanones *2a*, *2b*, and *3* were found to have mutagenic activities when tested with the *Salmonella typhimurium* strain TA 100.[12,13] When compared with other mutagens in spent chlorination liquor, activities were similar to 1,3-dichloroacetone but less than 2-chloropropenal[14] and much less than the potent mutagen 3-chloro-4-(dichloromethyl)-5-hydroxy-2(5H)-furanone[15] to the same *S. typhimurium* strain. R_m^o ("log P") values ranged from 2.6 for the isomeric furanones *2a* and *2b* to 3.4 for the furanone *3*.[12,13] The dione *1* has a lower R_m^o value but was difficult to separate completely from the more lipophilic material because of the large amount present.

Humic Acid

A large number of the chlorinated degradation products of lignin are also formed when humic acid is chlorinated. Thus, Coleman et al.[16] identified a number of chlorinated ketones, phenols, and thiophenes among chlorination products of humic acid and showed the by-products of chlorination of Cincinnati, Ohio, drinking water were similar to those of humic acid. Kringstad et al.[17,18] showed the same chloroacetones and chlorophenols found in spent bleach liquors were also formed when humic acid was chlorinated, and 3-chloro-4-(dichloromethyl)-5-hydroxy-2(5H)-furanone has been identified in drinking water.[19] It therefore seemed likely that the furanones *2a*, *2b*, and *3* would also be formed when humic acid was chlorinated.

A GC/MS chromatogram of the hexane extract from the chlorination of an unbuffered pH 7 solution of humic acid containing 1 g TOC/L and employing a Cl_2:C ratio (w/w) of 4:1 is shown in Figure 3. Tetrachlorocyclopentene-1,3-dione *1* and the furanone *2b* are seen to be prominent in the

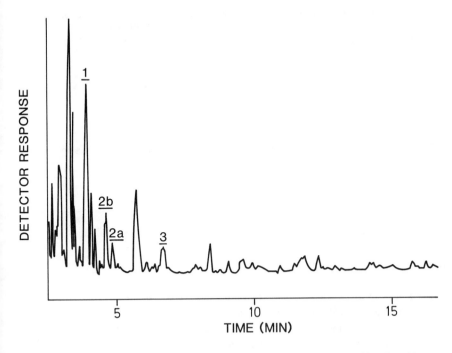

Figure 3.   Ion chromatogram of the hexane extract from the chlorination of humic acid.

chromatogram, whereas the furanones *2a* and *3* are minor components. This pattern is similar to that observed in the chlorination of lignin.

Traces of *1-3* were also detected by GC/ECD when humic acid was chlorinated under different pH conditions at a carbon concentration of 70 mg/L using a Cl_2:C ratio of 4:1 (Table 1). The effect of pH was most clearly seen in the case of the two major compounds *1* and *2b*. Chlorination at low pH for 1 hour yielded distinctly larger amounts of *1* and *2b* than when performed buffered at pH 7 then lowered to 1.5 just before extraction, or when performed at pH 1.5, raised to pH 7, then lowered again for extraction. Therefore, these compounds show similar pH susceptibility to many mutagenic chlorination products of lignin and humic acid.

Chemical Stability

The stability of compounds *1-3* is shown in Table 2. Although as mentioned earlier the dione *1* should not be considered lipophilic, it was included, since the presence of the furanone *3* is probably closely related (see next section). As can be seen from the results, the dione *1* disappeared completely within 1 hour at pH 7. This compound is known to be converted to the hydroxylactone *4* (see next section) under these conditions.[20] The

Table 1.   Quantities of Tetrachlorocyclopentene-1,3-dione *1* and Furanones *2a*, *2b*, and *3* Produced in the Chlorination of Humic Acid (ng/mg TOC)

pH	Compound			
	1	*2a*	*2b*	*3*
1.5	710	<20	250	<20
1.5→7→1.5	<20	95	60	20
7→1.5	40	95	135	<20

Table 2.   Stability of Tetrachlorocyclopentene-1,3-dione and Furanones in H$_2$O at pH 7

Compound	% Remaining				
	1 h	4 h	10 h	24 h	48 h
1	0	—	—	—	—
2a	86	69	77	36	35
2b	83	40	7	0	—
3	75	27	1	0	—

furanones *2b* and *3* disappeared in 24 hours but 35% of the furanone *2a* was still present after 48 hours. The greater stability of *2a* may result from the absence of an inductive effect by chlorine adjacent to the carbonyl group, which is the site of reaction with water.

Origin and Synthesis of *2a*, *2b*, and *3*

The furanones *2a*, *2b*, and *3* may originate from chlorocyclopentene-1,3-diones. Tetrachlorocyclopentene-1,3-dione *1* hydrolyzes readily to the hydroxylactone *4*, a constituent of spent chlorination liquor.[20] The hydrolysis is pH-dependent and proceeds more rapidly with increasing pH. The lactol *4* is dehydrated to the furanone *3* under strong acid conditions[21] as shown in Figure 4.

In order to determine whether the furanones *2a* and *2b* could be formed by analogous reactions of the unknown trichlorocyclopentene-1,3-dione *6*, acid cyclization of the unsaturated ester *5*[22] to the dione *6* was attempted (Figure 5) as described by Roedig and Märkl[21] for the fully chlorinated compound.

Attempts to purify the dione *6* revealed it was considerably more unstable than the analogous perchlorinated dione *1*, which is a crystalline compound. However, a product was obtained which had spectroscopic properties consistent with *6*. Further heating with acid of solutions containing *6* resulted in formation of the furanone *2b* in 14% overall yield from the ester *5* and a small amount of the furanone *2a*. These results confirmed the possible formation of the two trichlorofuranones *2a* and *2b* by an acid-mediated reaction via the dione *6*.

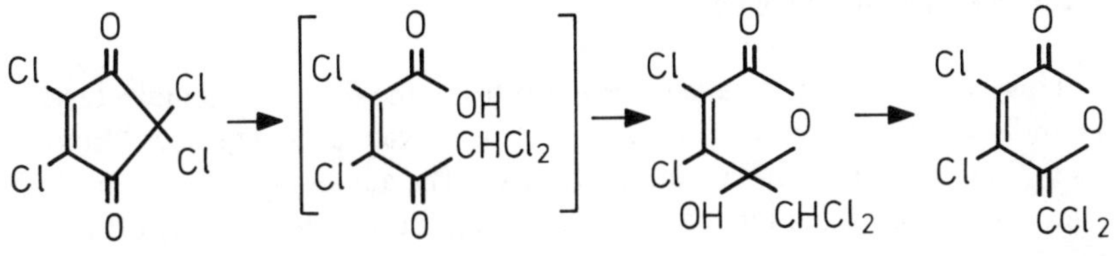

Figure 4. Conversion of tetrachlorocyclopentene-1,3-dione *1* to the furanone *3* via the intermediate lactol *4*.

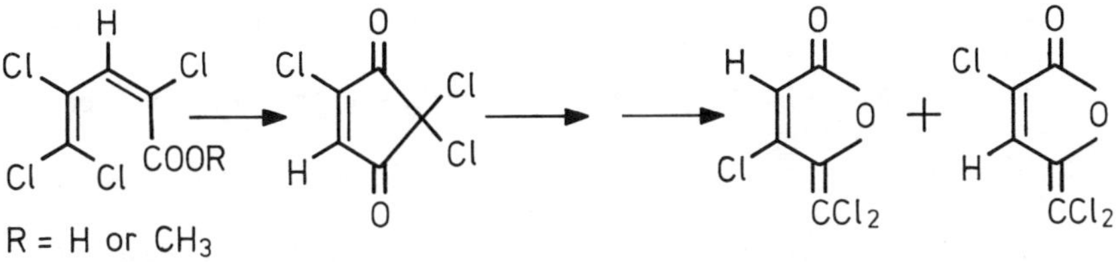

Figure 5. Synthesis of the furanones *2a* and *2b*.

Significance of Present Findings

The chlorinated furanones discussed in this chapter are formed only in trace amounts during the chlorination of softwood kraft pulp and humic acid. In the case of pulp bleaching, the indications are that even smaller quantities are produced when the kappa number of unbleached pulp is reduced by oxygen prebleaching prior to bleaching with chlorine. In the case of humic acid, the furanones were only detected readily when chlorinations were performed under favorable conditions, e.g., high carbon concentrations and low pH. Chemical stability tests also indicate these compounds are unstable at neutral pH. Although it is unlikely the furanones described constitute an environmental hazard in receiving waters, their significance with respect to drinking water is unknown.

In addition to the work described here, other lipophilic compounds produced by the chlorination of pulp lignin and humic acids are under investigation. These include the chlorinated thiophenes and cymenols, which have been reported in a variety of studies and have also been found in the lipophilic fractions recovered from RPTLC in our work. These compounds appear to be somewhat resistant to biodegradation.[23]

ACKNOWLEDGMENTS

This investigation received support from the Environmental Research Foundation of the Swedish Pulp and Paper Association, project "Environment 90—project 1—Bleaching Effluents." The authors received many valuable suggestions from Dr. Lars M. Strömberg during the course of the work. GC/MS work was done by Pierre Ljungquist, and mutagenicity and lipophilicity measurements by Marie-Claude Kolar.

REFERENCES

1. Leo, A., C. Hansch, and D. Elkins. "Partition Coefficients and Their Uses," *Chem. Rev.* 71:525–616 (1971).
2. *OECD Guidelines for Testing of Chemicals* (Paris: Organisation for Economic Co-operation and Development, 1981).
3. Kringstad, K. P., and K. Lindström. "Spent Liquors from Pulp Bleaching," *Environ. Sci. Technol.* 18:236A-248A (1984).
4. Landner, L., K. Lindström, M. Karlsson, J. Nordin, and L. Sörensen. "Bioaccumulation in Fish of Chlorinated Phenols from Kraft Pulp Mill Bleachery Effluents," *Bull. Environ. Contamin. Toxicol.* 18:663–673 (1977).
5. "Environmentally Harmonized Production of Bleached Pulp." Final Report. Swedish Forest Products Research Laboratory, Stockholm (1982).
6. McKague, A. B. "Some Toxic Constituents of Chlorination-Stage Effluents from Bleached Kraft Pulp Mills," *Can. J. Fish Aquat. Sci.* 38:739–743 (1981).
7. Lindström, K., and R. Schubert. "Determination by MS-MS of 1,1-Dichlorodimethyl Sulfone from Pulp Mill Bleach Plant Effluents in Aquatic Organisms," *J. High Resol. Chromatog. Chromatog. Commun.* 7:68–73 (1984).
8. The Swedish Pulp and Paper Association, Stockholm (1987).
9. Butte, W., C. Fooken, R. Klussmann, and D. Schuller. "Evaluation of Lipophilic Properties for a Series of Phenols using Reversed-Phase High-Performance Liquid Chromatography and High-Performance Thin-Layer Chromatography," *J. Chromatog.* 214:59–67 (1981).
10. Renberg, L. O., S. G. Sundström, and A-C. Rosén-Olofsson. "The Determination of Partition Coefficients of Organic Compounds in Technical Products and Waste Waters for the Estimation of their Bioaccumulation Potential using Reversed Phase Thin Layer Chromatography," *Toxicol. Environ. Chem.* 10:333–349 (1985), and references therein.
11. Kringstad, K. P., F. de Sousa, and L. M. Strömberg. "Evaluation of Lipophilic Properties of Mutagens Present in the Spent Chlorination Liquor from Pulp Bleaching," *Environ. Sci. Technol.* 18:200–203 (1984).
12. Kringstad, K. P., P. O. Ljungquist, A. B. McKague, F. de Sousa, and L. M. Strömberg. "Lipophilic Compounds in Spent Bleach Liquors," paper presented at the 4th International Symposium on Wood and Pulping Chemistry, Paris, April 27–30, 1987.

13. McKague, A. B., M-C. Kolar, and K. P. Kringstad. "Nature and Properties of Lipophilic Organic Compounds in Spent Liquors from Pulp Bleaching. Part I: Liquors from Conventional Bleaching of Softwood Kraft Pulp," *Environ. Sci. Technol.*, submitted for publication.

14. Bull, R. J., and M. Robinson. "Carcinogenic Activity of Haloacetonitrile and Haloacetone Derivatives in the Mouse Skin and Lung," in *Water Chlorination: Chemistry, Environmental Impact and Health Effects, Vol. 5*, R. L. Jolley, R. J. Bull, W. P. Davis, S. Katz, M. H. Roberts, Jr., and V. A. Jacobs, Eds. (Chelsea, MI: Lewis Publishers, Inc., 1985), 221.

15. Holmbom, B., R. H. Voss, R. D. Mortimer, and A. Wong. "Fractionation, Isolation, and Characterization of Ames Mutagenic Compounds in Kraft Chlorination Effluents," *Environ. Sci. Technol.* 18:333–337.

16. Coleman, W. E., J. W. Munch, W. H. Kaylor, R. P. Streicher, H. P. Ringhand, and J. R. Meier. "Gas Chromatography/Mass Spectrometry Analysis of Mutagenic Extracts of Aqueous Chlorinated Humic Acid. A Comparison of the Byproducts to Drinking Water Contaminants," *Environ. Sci. Technol.* 18:674–681 (1984).

17. Kringstad, K. P., F. de Sousa, and L. M. Strömberg. "Studies on the Chlorination of Chlorolignins and Humic Acid," *Environ. Sci. Technol.* 19:427–431 (1985)

18. Kringstad, K. P., P. O. Ljungquist, F. de Sousa, and L. M. Strömberg. "On the Formation of Mutagens in the Chlorination of Humic Acid," *Environ. Sci. Technol.* 17:553–555 (1983).

19. Hemming, J., B. Holmbom, M. Reunanen, and L. Kronberg. "Determination of the Strong Mutagen 3-Chloro-4-(dichloromethyl)-5-hydroxy-2(5H)-furanone in Chlorinated Drinking and Humic Waters," *Chemosphere* 15:549–556 (1986).

20. Strömberg, L. M., F. de Sousa, P. Ljunquist, B. McKague, and K. P. Kringstad. "An Abundant Chlorinated Furanone in the Spent Chlorination Liquor from Pulp Bleaching," *Environ. Sci. Technol.* 21:754–756 (1987).

21. Roedig, A., and G. Märkl. "Constitution of Different $Cl_5Cl_4O_2$ Compounds. Perchloro Derivatives of 1-Cyclopentene-3,5-dione, 2-pyrone, and protoanemonin," *Liebigs Ann. Chem.* 636:1–18 (1960).

22. Märkl, G. "Synthesis of α-halogenated, α,β-unsaturated carboxylic acids with triphenylphosphine-halocarbomethoxy-methylenes," *Chem. Ber.* 94:2996–3004 (1961).

23. Voss, R. H. "Neutral Organic Compounds in Biologically Treated Bleached Kraft Mill Effluents," *Environ. Sci. Technol.* 18:938–946 (1984).

CHAPTER 11

Amino Acids as Model Compounds for Halogenated By-Products Formed on Chlorination of Natural Waters

Michael L. Trehy, Monsanto Company, St. Louis, Missouri

Richard A. Yost, Chemistry Department, University of Florida, Gainesville, Florida

Carl J. Miles,* Pesticide Research Laboratory, University of Florida, Gainesville, Florida

INTRODUCTION

The presence of trihalomethanes in drinking water disinfected with chlorine was revealed by Bellar et al. in 1974.[1] Humic and fulvic substances containing resorcinol-type moieties were determined by Rook to be likely precursors for trihalomethanes.[2] Further work by Symons et al.[3] and by Christman et al.[4] confirmed that the major source for trihalomethanes in drinking water could be explained by the reaction of humic and fulvic substances with chlorine. The detection of dichloroacetonitrile in drinking water was reported by McKinney et al.[5] and by Stevens,[6] although the source for dichloroacetonitrile in the samples was not determined. Resorcinol-type moieties cannot account for the formation of dichloroacetonitrile in chlorinated natural waters. The formation of dihaloacetonitriles (DHAN) by the reaction of aqueous chlorine with amino acids or other nitrogenous compounds with amino acid moieties was proposed by Trehy[7] and Bieber.[8] Certain amino acids such as aspartic acid, tyrosine, and tryptophan yield DHAN and chloral hydrate when exposed to aqueous chlorine under conditions similar to that used in water treatment.[9]

---

*Current affiliation: University of Hawaii at Manoa, Honolulu, Hawaii.

RESULTS AND DISCUSSION

Reaction of Amino Acids with Aqueous Chlorine

Studies regarding the chlorination of amino acids and peptides have shown that the primary amino group on the amino acid or peptide can be converted to either an aldehyde or nitrile group.[10-20] These studies indicate that with an equimolar amount of halogenating agent the major product is the aldehyde. However, if an excess of halogenating agent is added, then the corresponding nitrile can also be formed, with the ratio of the aldehyde to nitrile formed increasing with pH.[15] The reaction pathway for amino acids involves initial rapid formation of the mono- and dichloroamine, which can react further to form the aldehyde or nitrile, respectively. A reaction pathway suggested by Friedman and Morgulis[14] is shown in Figure 1. For alanine, R is a methyl group and acetaldehyde and acetonitrile are formed on chlorination. The presence of unhalogenated aldehydes and nitriles in chlorinated natural waters[21] can be attributed in part to the presence of amino acids or peptides in natural waters. Humic acids may also contribute to the presence of amino acids in natural waters. Humic acids may have amino acids associated with them either in a free or in a combined form. The formation of 3-cyanopropanoic acid (CPA) and 4-cyanobutanoic acid (CBA) on chlorination of humic acids has been demonstrated by de Leer et al.[22] The amino acids glutamic acid and lysine were suggested by de Leer et al. as possible precursors, since CPA and CBA were found to be formed on chlorination of these amino acids.[23]

Formation of Dichloroacetonitrile and Chloral Hydrate

Aspartic acid, tyrosine, and tryptophan react with aqueous chlorine to yield dichloroacetonitrile, chloral hydrate, and trichloromethane.[9] The reaction pathway for these amino acids is quite complicated, with many intermediates and by-products formed. In the case of aspartic acid, de Leer demonstrated the presence of at least 11 other significant by-products.[23] The major reaction products for amino acids with chlorine are oxidation by-products. However, when the amino acid contains an additional activating group near the alpha amino acid functional group, then a substitution reaction may readily occur. The presence of the second carboxylic acid group in aspartic acid increases the reactivity of the beta hydrogens to chlorination. A simplified pathway for aspartic acid reacting with aqueous chlorine is shown in Figure 2. Aspartic acid reacts with aqueous chlorine in a manner very similar to the general pathway suggested by Friedman and Morgulis.[14] As the pH at which a sample of aspartic acid was chlorinated was raised from 6.4 to 8.5, the ratio of dichloroacetonitrile to chloral hydrate formed rose from 0.6 to 9. Aspartic acid was found to yield 5 to 10% dichloroacetonitrile and chloral

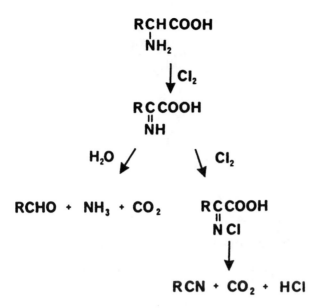

Figure 1. Pathway proposed by Friedman and Morgulis for chlorination of amino acids.[14]

hydrate on a molar basis in approximately 1 to 1.5 hours. The relatively low yield may be due in part to the formation of numerous other reaction products.[23] Chloroform was not formed during the chlorination step but appears to result entirely as a result of the relatively slow hydrolysis of chloral. The half-life of chloral at pH 8 and 35°C is 2 days.[24]

Tyrosine and tryptophan yield trichloromethane, dichloroacetonitrile, and chloral hydrate immediately after chlorination, suggesting that chlorination can occur at more than one site, since trichloromethane was not formed in the reaction with aspartic acid. Chlorination of tyrosine at pH 1–2[9,16,18] has been found to result in the formation of mono- and dichloro-4-hydroxyphenylacetonitrile and mono- and dichloro-4-hydroxyphenylacetaldehyde. The results of chlorination studies at low pH clearly point to chlorination of the aromatic ring, which could ultimately result in the formation of trichloromethane. Chlorination of the amino acid side chain would be expected to result in the formation of dichloroacetonitrile and chloral hydrate, as shown in Figure 3 for tyrosine. Tryptophan reacts to form simple chlorination byproducts much more rapidly than does tyrosine. In 1 hour at pH 8.5, a 45.6 ppm solution of tyrosine yielded 0.31 ppm trichloromethane, 0.019 ppm dichloroacetonitrile, and 0.23 ppm chloral hydrate, while a 58.4 ppm solution of tryptophan yielded 5.2 ppm trichloromethane, 0.23 ppm dichloroacetonitrile, and 6.2 ppm chloral hydrate.

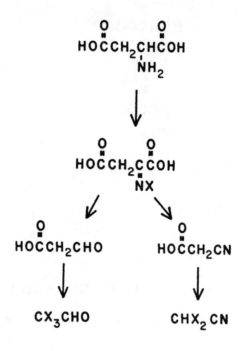

Figure 2. Chlorination of aspartic acid.

Determination of Dihaloacetonitriles (DHAN) and Chloral Hydrate

Detection and quantification of DHAN and chloral in chlorinated natural waters is complicated by (a) hydrolysis of DHAN and chloral to dihaloacetic acids and chloroform, respectively,[7,8] (b) degradation of DHAN by dechlorinating agents such as sodium sulfite and sodium thiosulfate,[7,8] (c) low purge efficiency for the DHAN and chloral in the purge-and-trap technique,[9] and (d) low extraction efficiency for chloral with pentane in the liquid-liquid extraction procedure normally used.[7-9] Although chloral is not efficiently extracted from water with pentane, chloral can be extracted with an efficiency of approximately 36% when the ratio by volume of ethyl ether to water is 1:5. Ethyl ether quantitatively extracts dichloroacetonitrile and trichloromethane, permitting simultaneous analysis for all three by-products. Chloral hydrate was determined in this study by gas chromatography/mass spectrometry with a diethylene glycol succinate-packed column. Detection of low ppb aqueous concentrations of chloral and dichloroacetonitrile is accomplished by negative chemical ionization-selected ion monitoring mass spectrometry. However, Hodgeson and Cohen have developed a significantly better method, employing a DB-1 capillary column.[25] Chloral hydrate

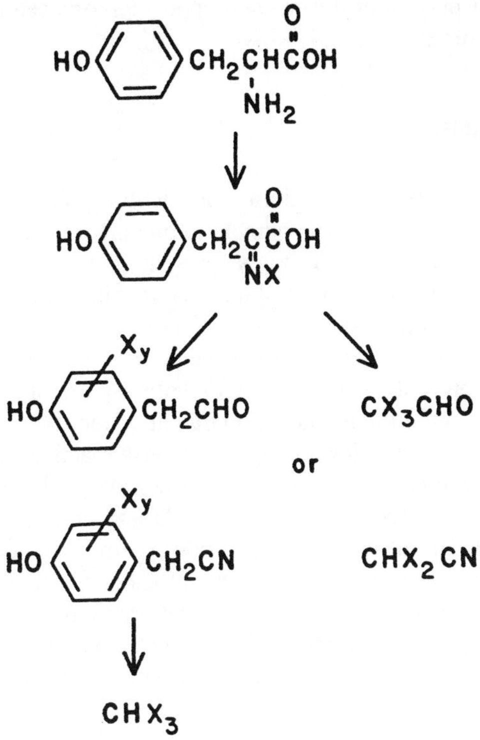

Figure 3.   Chlorination of tyrosine.

decomposes on the packed column to trichloroacetaldehyde, resulting in considerable band broadening, which does not appear to be a significant problem with the DB-1 capillary column.

Chlorinated Natural Waters

Lake water samples were chlorinated in order to demonstrate the formation of both dichloroacetonitrile and chloral hydrate. A sample of lake water from West Palm Beach, Florida, was chlorinated at pH 7.2 for 70 minutes and produced 239 ppb of chloroform, 19 ppb of dichloroacetonitrile, and 30 ppb of chloral hydrate, while a lake water sample from Gainesville, Florida, chlorinated at pH 7.6 yielded 116 ppb of chloroform, 17 ppb of dichloroacetonitrile, and 35 ppb of chloral after 70 minutes. Although the relative concentrations of dichloroacetonitrile and chloral hydrate were similar, the chloroform concentration was substantially different, possibly reflecting different precursors for these by-products. Chlorinated wastewater from an extended aeration wastewater treatment plant in Gainesville, Florida, was found to contain 32–80 ppb chloroform, 7–14 ppb dichloroacetonitrile, and 20–38 ppb chloral hydrate.

CONCLUSIONS

Although the presence of chloral and dichloroacetonitrile in the chlorinated water samples may be attributable to precursors other than amino acids, the potential for amino acids to be present in the lake water and in the wastewater is well documented. Amino acids either in a combined form or possibly as free amino acids are common constituents of the environment.[26-30] Coastal plain rivers of Southeastern United States contain approximately 2 ppm of total amino acids with aspartic acid at approximately 0.3 ppm.[27] Although the amino acid content of the rivers is thought to exist primarily in a combined form that may be less reactive than the free amino acids, the environmental concentration of amino acids appears to be sufficient to account for the dichloroacetonitrile and chloral hydrate formed on chlorination of natural waters. In addition to amino acids, peptides present either in the environment or in vivo can also react with aqueous chlorine to form N-chloroamines.[13,16,31] The chlorination by-products of amino acids are of concern, because amino acids are likely to be present in the environment and a portion of the by-products formed on chlorination may be mutagenic.[32-33]

REFERENCES

1. Bellar, T. A., J. J. Lichtenberg, and R. C. Kroner. "The Occurrence of Organohalides in Chlorinated Drinking Waters," *J. Am. Water Works Assoc.* 23:703–706 (1974).
2. Rook, J. J. "Formation of Haloforms During Chlorination of Natural Waters," *J. Water Treat. Exam.* 23:234–243 (1974).
3. Symons, J. M., T. A. Bellar, J. Carswell, J. DeMarco, K. L. Kropp, G. C. Robeck, D. R. Seeger, C. J. Slocum, B. C. Smith, and A. A. Stevens. "National Organics Reconnaissance Survey for Halogenated Organics," *J. Am. Water Works Assoc.* 67:634–647 (1975).
4. Christman, R. F., D. L. Norwood, D. S. Millington, J. D. Johnson, and A. A. Stevens. "Identity and Yields of Major Halogenated Products of Aquatic Fulvic Acid Chlorination," *Environ. Sci. Technol.* 17:625–628 (1983).
5. McKinney, J. D., R. R. Maurer, J. R. Hass, R. O. Thomas. "Possible Factors in the Drinking Water of Laboratory Animals Causing Reproductive Failure," in *Identification and Analysis of Organic Pollutants in Water*, L. H. Keith, Ed. (Ann Arbor, MI: Ann Arbor Science Publishers, Inc., 1976), 417–432.
6. Stevens, A. A. "Formation of Non-Polar Organo-Chloro Compounds as Byproducts of Chlorination," in *Proceedings—Oxidation Techniques in Drinking*

Water Treatment, September 11–13, 1978, Karlsruhe, F.R.G., EPA-570/9–79020, USEPA, Washington, DC (1979), pp. 145–160, NTIS Accession No. PB 301313/AS.

7. Trehy, M. L., and T. I. Bieber. "Detection, Identification and Quantitative Analysis of Dihaloacetonitriles in Chlorinated Natural Waters," in *Advances in the Identification and Analysis of Organic Pollutants in Water, Vol. 2*, L. H. Keith, Ed. (Ann Arbor, MI: Ann Arbor Science Publishers, Inc., 1981), 941–975.

8. Bieber, T. I., and M. L. Trehy. "Dihaloacetonitriles in Chlorinated Waters," in *Water Chlorination: Environmental Impact and Health Effects, Vol. 4*, R. L. Jolley, W. A. Brungs, J. A. Cotruvo, R. B. Cumming, J. S. Mattice, and V. A. Jacobs, Eds. (Ann Arbor, MI: Ann Arbor Science Publishers, Inc., 1983), 85–96.

9. Trehy, M. L., R. A. Yost, and C. J. Miles. "Chlorination Byproducts of Amino Acids in Natural Waters," *Environ. Sci. Technol.* 20:1117–1122 (1986).

10. Langheld, K. "Über den Abbau der α-Aminosauren zu fetten Aldehyden mittels Natriumhypochlorit," *Ber. Deutsch. Chem. Ges.* 42:392 (1909).

11. Dakin, H. D. "The Oxidation of Amino-Acids to Cyanides," *Biochem. J.* 10:319–323 (1916).

12. Dakin, H. D. "On the Oxidation of Amino-Acids and of Related Substances with Chloramine-T," *Biochem. J.* 11:79–95 (1917).

13. Goldschmidt, S., E. Wilberg, F. Nagel, and K. Martin. "Über Protein IV," *Justus Liebigs Ann. Chem.* 456:1–38 (1927).

14. Friedman, A. H., and S. Morgulis. "The Oxidation of Amino Acids with Sodium Hypobromite," *J. Am. Chem. Soc.* 58:909-913 (1936).

15. Morris, J. C., N. Ram, B. Baum, and E. Wajon. "Formation and Significance of N-Chloro Compounds in Water Supplies," EPA-600/2–80–031 (Cincinnati: U.S. Environmental Protection Agency, 1980).

16. Pereira, W. E., Y. Hoyano, R. E. Summons, V. A. Bacon, and A. M. Duffield. "Chlorination Studies. II. The Reaction of Aqueous Hypochlorous Acid with α-Amino Acids and Dipeptides," *Biochimica et Biophysica Acta* 313:170–180 (1973).

17. Stanbro, W. D., and W. D. Smith. "Kinetics and Mechanism of the Decomposition of N-Chloroalanine in Aqueous Solution," *Environ. Sci. Technol.* 13:446–451 (1979).

18. Burleson, J. L., G. R. Peyton, and W. H. Glaze. "Gas-Chromatographic/Mass-Spectrometric Analysis of Derivatized Amino Acids in Municipal Wastewater Products," *Environ. Sci. Technol.* 14:1354–1359 (1980).

19. Fox, S. W., and M. W. Bullock. "Synthesis of Indoleacetic Acid from Glutamic Acid and a Proposed Mechanism for the Conversion," *J. Am. Chem. Soc.* 73:2754–2755 (1951).

20. Isaac, R. A., and J. C. Morris. "Rates of Transfer of Active Chlorine Between Nitrogenous Substrates," in *Water Chlorination: Environmental Impact and Health Effects, Vol. 3*, R. L. Jolley, W. A. Brungs, and R. B. Cumming, Eds. (Ann Arbor, MI: Ann Arbor Science Publishers, Inc., 1980), 183–191.

21. Coleman, W. E., R. D. Lingg, R. G. Melton, and F. C. Kopfler. "The Occurrence of Volatile Organics in Five Drinking Water Supplies Using Gas Chromatography/Mass Spectrometry," in *Identification and Analysis of Organic Pollut-*

ants in Water, L. H. Keith, Ed. (Ann Arbor, MI: Ann Arbor Science Publishers, Inc., 1976), 305.

22. De Leer, E. W. B., J. S. Sinninghe Damste, C. Erkelens, and L. De Galan. "Identification of Intermediates Leading to Chloroform and C-4 Diacids in the Chlorination of Humic Acids," *Environ. Sci. Technol.* 19:512–522 (1985).

23. De Leer, E. W. B., T. Baggerman, P. van Schaik, C. W. S. Zuydeweg, and L. De Galan. "Chlorination of ω-Cyanoalkanoic Acids in Aqueous Medium," *Environ. Sci. Technol.* 20:1218-1223 (1986).

24. Luknitskii, F. I. "The Chemistry of Chloral," *Chem. Rev.* 75:259–289 (1975).

25. Hodgeson, J., and C. Cohen. "Analysis of Chlorinated By-Products in Drinking Water," preprint extended abstract, American Chemical Society Environmental Chemistry Division 27(2):479–482 (1987).

26. Peake, E., B. L. Baker, and Hodgson, G. W. "Hydrogeochemistry of the Surface Waters of the Mackenzie River Drainage Basin, Canada. II. The Contribution of Amino Acids, Hydrocarbons and Chlorines to the Beaufort Sea by the Mackenzie River System," *Geochim. Cosmochim. Acta* 36:867–883 (1972).

27. Beck, K. G., J. H. Reuter, and E. M. Perdue. "Organic and Inorganic Geochemistry of Some Coastal Plain Rivers of the Southeastern United States," *Geochim. Cosmochim. Acta* 38:341–364 (1974).

28. Lytle, C. R., and E. M. Perdue. "Free, Proteinaceous, and Humic-Bound Amino Acids in River Water Containing High Concentrations of Aquatic Humus," *Environ. Sci. Technol.* 15:224–228 (1981).

29. Sowden, F. J., and K. C. Ivarson. "The 'Free' Amino Acids of Soil," *Can. J. Soil Sci.* 46:109–120 (1966).

30. Hunter, J. V. "Origin of Organics from Artificial Contamination," in *Organic Compounds in Aquatic Environments,* S. D. Faust and J. V. Hunter, Eds. (New York: Marcel Dekker, 1971), 51–94.

31. Scully, F. E., Jr., and K. Mazina. "Toxicological Significance of the Chemical Reactions of Aqueous Chlorine and Chloramines," preprint extended abstract, American Chemical Society Environmental Chemistry Division 27(2):585–586 (1987).

32. Simmon, V. F., K. Kauhanen, and R. G. Tardiff. "Mutagenic Activity of Chemicals Identified in Drinking Water," *Dev. Toxicol. Environ. Sci.* 2:249–258 (1977).

33. Bull, R. J. "Health Risks of Drinking Water Disinfectants and Disinfection By-Products," in *Water Chlorination: Environmental Impact and Health Effects, Vol. 4,* R. L. Jolley, W. A. Brungs, J. A. Cotruvo, R. B. Cumming, J. S. Mattice, and V. A. Jacobs, Eds. (Ann Arbor, MI: Ann Arbor Science Publishers, Inc., 1983), 1401–1415.

CHAPTER 12

Toxicological Significance of the Chemical Reactions of Aqueous Chlorine and Chloramines

Frank E. Scully, Jr., Kathryn Mazina, and Daniel E. Sonenshine, Department of Chemical Sciences, Old Dominion University, Norfolk, Virginia

H. Paul Ringhand, U.S. Environmental Protection Agency, Cincinnati, Ohio

INTRODUCTION

When a strong oxidizing agent (e.g., a drinking water disinfectant) is ingested, it dissolves in the highly organic solutions of the gastrointestinal tract, including saliva and stomach fluid, and comes in contact with tissues. Because it is a strong oxidant, it can be expected to react with organic biomolecules present. Since aqueous chlorine and, to an ever-increasing extent, inorganic monochloramine are widely used disinfectants, it is important to determine what reactions these oxidants can undergo in the body and whether their products exhibit adverse health effects or can be detoxified. In this chapter, the reactions of aqueous chlorine and monochloramine with biomolecules and in biological systems will be reviewed with the intent of assessing their toxicological significance.

Several considerations make the study of the reactions of hypochlorite in the gastrointestinal tract more complex than reactions in natural waters. First, the stomach is a difficult reaction vessel in which to study chemical reactions. It is like a beaker with a hole in it. It leaks! As a result, it holds solids much better than it does liquids so that there is rarely a quantitative recovery of reagents added to an active stomach. Second, the stomach is a more dynamic system than that usually studied by environmental chemists. In a hungry person or animal, the sight or smell of food induces the secretion of copious amounts of gastric fluid, which is strongly acidic and rich in the digestive enzyme pepsin and in mucin, a collection of glycoproteins.[1] The resulting concentration of these organic carbon compounds is several orders of magnitude greater than that found in natural waters and therefore the chlorine-carbon reaction ratio is much smaller in gastric fluid after ingestion of chlorinated water than that found in a drinking water treatment system. For this reason, the products of the reactions of aqueous chlorine in gastric fluid are not likely to be as extensively chlorinated or oxidized as are the by-products of drinking water chlorination like chloroform and trichloroacetic acid. This also means that identification of chlorinated products formed in this medium will be more difficult because the by-products may not be small fragments of the chlorination of larger molecules.

The types of organic compounds present in stomach fluid depend on diet and include all major classes of biomolecules, such as proteins, polysaccharides, lipids, and vitamins. A full bland diet adds approximately 100 g protein, 125 g fat, and 275 g carbohydrate per day to the organic composition of gastric fluid.[2] Third, the pH ranges from a low of about 1.3 in an actively digesting stomach to approximately neutral pH in saliva[1] to pH 7.8–8.0 in pancreatic juice.[3] In the stomach alone, the pH might fluctuate from that of the solution ingested, such as pH 8 for tap water, to pH 2 shortly after eating begins, when hydrochloric acid is excreted from parietal cells in the stomach. At these low pH values, the ionic strength is also much higher than that of natural waters.

Many pharmacokinetic studies are conducted with fasted animals. Of particular interest to studies involving laboratory rats is the observation that fasted animals, stimulated by the taste of food, excrete approximately 0.8 mL of gastric fluid per hour containing a hydrogen ion concentration of approximately 65 mEq/L.[4] The pH of this solution alone would be approximately 1.2. However, it would be overly simplistic to describe the contents of the fluid in the stomach of a fasted animal by an analysis of gastric secretions. We have observed that even in fasted animals partially digested food is still present in the stomach 48 hr after fasting is begun. Consequently, a solution of an oxidant orally administered to a fasting animal encounters a complex mixture of organic compounds derived from partially digested food and gastric secretion.

With these factors in mind, the reactions of aqueous chlorine and monochloramine (NH_2Cl) in the stomach fluid of laboratory rats have been addressed. This work has been restricted to studies of the reactions in the gastric fluid from fasted animals to mimic conditions encountered in pharmacokinetic studies and to minimize the variables added by the organic components of excess food interacting with the disinfectant. Fukayama et al.[5] have reviewed the reactions of chlorine with the major classes of biomolecules that are found in foods. Only proteins, amino acids, and unsaturated lipids react with aqueous chlorine to form chlorinated compounds, while carbohydrates form unchlorinated oxidation products. Because organic nitrogen compounds are major organic components of gastric fluid as well as important dietary constituents, much attention has been focused on the products of the reactions of these compounds with aqueous chlorine and monochloramine.

ORGANIC NITROGEN COMPONENTS OF STOMACH FLUID

The concentrations of proteins and amino acids in gastric fluid isolated from rats fasted 48 hr and administered 4 mL deionized water have been measured. The concentrations of total Kjeldahl nitrogen (TKN) ranged from 200 to 400 mg/L (as N)[6] and protein nitrogen concentrations measured by the biuret method[7] accounted for 70–99% of the TKN.[6] The sum of the amino acid concentrations in the stomach fluid of animals fasted 48 hr and administered 3 mL deionized water averaged 204 μM or about 1% of the TKN.[8]

REACTIONS OF HOCl IN GASTRIC FLUID

If a complex mixture of reagents reacts with aqueous chlorine and either reduces it or converts it to a different oxidizing species, it exhibits a "chlorine demand." Usually this term is used in reference to natural waters or wastewaters to determine the amount of oxidant that must be added to obtain a residual chlorine level after a given time. The chlorine demand of gastric fluid recovered from animals administered 4 mL of deionized water was measured by diluting the fluid, chlorinating it, measuring the residual chlorine level, and correcting the residual measured for the dilution of the fluid. When the fluid from animals fasted 48 hr was chlorinated to concentrations ranging from 100 to 600 mg/L (as Cl_2), between 40% and 80% of the administered oxidant concentration remained after 15 min and was detected as combined residual chlorine.[6] This suggested that N-chlorinated amino acids and proteins were formed in this dosage range. At higher aqueous chlorine concentrations a free available chlorine was measured in addition to a stable combined residual chlorine. The formation of organic N-chloramines in stomach fluid is to be anticipated, since hypochlorous acid reacts very rapidly[9] and quantitatively[10] with organic amines. However, the rate of chloramine formation is dependent on the concentration of free amine and is therefore affected by pH, decreasing by a factor of 10 for each pH unit below approximately 7.0.

To determine what organic chloramines might form on administration of HOCl, gastric fluid was mixed with concentrated hypochlorous acid to a final concentration of 690 mg/L (as Cl_2) and derivatized with dansyl sulfinic acid. Dansyl sulfinic acid reacts with chloramines to form stable sulfonamide derivatives.[11] The derivatized mixtures were chromatographed by HPLC and derivatives of N-chloroalanine, N-chloroglycine, and N-chlorophenylalanine were tentatively identified by correlation of their retention times with known standards.[6]

Organic N-chloramines were also shown to form in vivo following successive administration of an amine (piperidine) or an amino acid (glycine) and hypochlorous acid.[12] ^{14}C-N-chloroglycine was identified after derivatization with dansyl sulfinic acid and ^{3}H-N-chloropiperidine was identified by direct HPLC analysis and comparison of its retention time with an authentic sample.

[3]H-Labeled piperidine was used to measure the yield of chloramines which would form in vivo in the stomach.[6] After successive administration (by gavage) of aqueous solutions of the radiolabeled piperidine and hypochlorite (200 mg/L and 1000 mg/L as Cl_2), the stomach contents were recovered. After correcting the yields of the chloramine for the chlorine demand of other components of the gastric fluid which would compete with the amine for the chlorine, the yield of tritiated N-chloropiperidine isolated accounted for 42% (1000 mg/L solution) and 70% (200 mg/L solution) of the estimated amount of chloramine which could form in the stomach fluid.

Mink et al.[13] have shown that chloroform, dichloroacetonitrile (DCAN), and di- and trichloroacetic acids can be formed in vivo following oral administration of 7 mL of a 8 mg/mL solution of sodium hypochlorite. Dichloroacetic acid and, in one instance, chloroform were found in plasma. DCAN can be formed by extensive chlorination and oxidation of aspartic acid, tyrosine, and tryptophan.[14] N-Chlorinated intermediates are involved.[15] Chloroform can also form by chlorination of proteins[16] and amino acids.[17] Since chloroform is not produced in natural waters disinfected by chloramines,[18] it must be assumed that hypochlorous acid was responsible for the formation of these compounds. However, as discussed above, hypochlorous acid was not detected in stomach fluid 15 min after animals were administered 4 mL of a solution containing less than 600 mg/L as Cl_2.[6] This suggests that the significance of the products reported by Mink et al.[13] would be much less at chlorine concentrations below 600 mg/L.

REACTIONS OF CHLORAMINES WITH ORGANIC COMPONENTS OF GASTRIC FLUID

Organic chloramino acids can potentially form in gastric fluid when it is mixed with inorganic monochloramine. In the absence of reducing reactions in the stomach, monochloramine can transfer its chlorine atom to organic amines and amino acids. Using the rate constants of Isaac and Morris[19,20] for the slower (by 10) of the two mechanisms which they propose and the concentrations of amino acids in gastric fluid reported above, half the chlorine in a solution of monochloramine with a concentration of 2 mg/L (as Cl_2) would be transferred to amino acids in about 9 min at pH 7.

Under conditions designed to simulate the gastrointestinal tract, Bercz and Bawa[21] demonstrated that monochloramine caused binding of radiolabeled iodide to nutrient biochemicals such as tyrosine, 4-aminobenzoic acid, arachidonic acid, and folic acid. Monochloramine also caused binding of iodide to organic components of saliva and gastric juice. The amount of binding varied with pH, but was generally less at neutral pH than at higher pH values.

In some of our work we have used N-chloropiperidine as a model to mimic the chemistry of the chlorinating oxidants HOCl and NH_2Cl. There are several very practical reasons for using a model compound instead of the actual disinfectant. First, inorganic monochloramine undergoes several reactions that would mask chemical studies of the compound in a complex mixture. For instance, it undergoes chlorine transfer from one monochloramine molecule to another (disproportionation) to form dichloramine and ammonia, the rate of which increases with decreasing pH below 8.[22] As discussed above, monochloramine also readily transfers its chlorine at a comparatively rapid rate to organic amines.[19,20] Therefore, the specific reactions of monochloramine are diminished by the loss of chlorine in the formation of other oxidants, e.g., dichloramine and organic chloramines. To inhibit these reactions, solutions of monochloramine are usually buffered above pH 8, but secretion of acid in the stomach would presumably overcome the inhibitory effects of buffers. It is difficult to assess the toxicological effects of monochloramine without knowing what the reacting species is. By contrast, N-chloropiperidine contains only one reactive chlorine atom and cannot form a dichloramine. Therefore, the chemistry of N-chloropiperidine is considerably cleaner than that of monochloramine. Second, N-chloropiperidine can be separated from other organic compounds and analyzed by high-performance liquid chromatography (HPLC).[12] As a result, the fate of a radiolabeled analogue of N-chloropiperidine in biological fluids can be traced easily. As far as is known, this is not possible with NH_2Cl, since no methods for the chromatographic separation of NH_2Cl have been reported.

When ^{36}Cl-N-chloropiperidine (final concentration of 300 mg/L as Cl_2) was dissolved at pH 7–8 in gastric fluid from rats fasted 48 hr and the mixture was incubated for 30 min at 37°C, 31% of the chloramine was reduced to chloride. However, 36% of the ^{36}Cl-labeled chlorine became bound to organic components of the stomach fluid,[23] while unreacted ^{36}Cl-N-chloropiperidine constituted the remaining 33% of the ^{36}Cl-label. The new ^{36}Cl-labeled material was not believed to be a single compound, but a group of polar, chlorinated compounds and is therefore referred to as a ^{36}Cl-chloroorganic fraction. Since the material did not oxidize iodide to iodine, the chlorine did not appear to be bonded to nitrogen, and was therefore thought to be bonded to carbon. Further work is currently under way to characterize these compounds. However, these preliminary results suggest that chloramines may react with organic components of stomach fluid to form new compounds with unknown health effects. It is hoped that results observed at active chlorine concentrations of 300 mg/L can be extrapolated to concentrations found in actual drinking waters, so that a realistic assessment of the health effects of these reactions can be made.

PHARMACOKINETIC STUDIES

Pharmacokinetic studies are used to determine the rate and extent a potentially toxic agent is absorbed, whether it is localized in a target organ, how it becomes metabolized, and how fast it is eliminated. However, for several reasons studies involving hypochlorous acid and inorganic monochloramine are not amenable to a "classical" interpretation given to pharmaceuticals. For instance, the speciation of the oxidant in aqueous solutions of chlorine is sensitive to pH. At pH values above 7.5, ClO^- is the reactive species, while at lower pH values HOCl predominates. At low pH and high chloride concentrations aqueous Cl_2 exists.[24] Monochloramine solutions are normally kept at pH values above 8, because as the pH is lowered its conversion to dichloramine is accelerated.[22]

Because aqueous chlorine and monochloramine are potent oxidants, pharmacokinetic studies of the radiolabel of $HO^{36}Cl$ or $NH_2^{36}Cl$ in animals do not reveal what happens to the parent compound, but rather to the product of the reactions of these compounds in vivo. Studies have suggested that the radioisotope is eliminated from fasted rats administered $HO^{36}Cl$ at about the same rate as from those administered ^{36}Cl-chloride.[25,26] Only ^{36}Cl-chloride was identified in an analysis of the plasma 120 hr after administration of $HO^{36}Cl$.[26] This would suggest that the hypochlorous acid is simply reduced to chloride. However, the normal concentration of chloride in human saliva is 45 mEq/L and its concentration in gastric fluid is as high as 170 mEq/L.[1] Concentrations are presumably as high in similar fluids from rats. Since $HO^{36}Cl$ undergoes rapid isotope exchange with unlabeled chloride (equation 1),[27]

$$HO^{36}Cl + Cl^- \rightleftharpoons HOCl + {}^{36}Cl^- \tag{1}$$

it is not clear whether the similarity between the elimination kinetics of $HO^{36}Cl$ and ^{36}Cl-chloride is due to complete reduction of the hypochlorous acid or to isotope exchange followed by elimination of chloride. However, it is worth noting that Abdel-Rahman et al.[26] found that when $HO^{36}Cl$ was administered to nonfasted animals, the ^{36}Cl activity was eliminated at a rate half that found with fasted animals. Without further information it is tempting to speculate that organic components of food reacted with the oxidant before isotope exchange could take place to form chlorinated products that were eliminated at a slower rate than ^{36}Cl-chloride. No studies of the corresponding isotope exchange of $NH_2^{36}Cl$ with chloride have been reported, but ^{36}Cl-N-chloropiperidine does not undergo significant isotope exchange over two hours from pH 2.5 to pH 8.[8]

In a pharmacokinetic study of $NH_2{}^{36}Cl$ in Sprague-Dawley rats, Abdel-Rahman et al.[26] observed kinetics for the absorption and elimination of the radiolabel which was markedly different from that observed with ^{36}Cl-chloride or $HO^{36}Cl$. The apparent concentration of radiolabel in plasma reached a maximum in about 8 hr and maintained a plateau for the next 40 hr. As a result, an average of 63% of the radiolabel was retained 48 hr after dosing by the rats administered $NH_2{}^{36}Cl$ compared with an average of 40% retained by rats administered ^{36}Cl-chloride.[25] If $NH_2{}^{36}Cl$ reacted with organic components of stomach fluid to form a ^{36}Cl-chlorinated organic fraction like that found when ^{36}Cl-N-chloropiperidine is mixed in vitro with gastric fluid from fasted animals,[23] this might account for the longer retention of the radiolabel. Further work is in progress to determine if there is any correlation between these observations.

ABSORPTION OF CHLORAMINES INTO BLOOD

Although chloramines are oxidizing agents, at least one is capable of being absorbed intact into blood.[12] When ^3H-N-chloropiperidine (1.85 mL of a solution with a concentration of 1784 mg/L as Cl_2, 3.3 mg as Cl_2 per animal) was administered to fasted rats, between 1% and 3% of the radioactivity appeared in the blood at intervals of 30, 60, and 120 min after administration. Most of the radioactivity was due to ^3H-piperidine, the product of the loss of the chlorine atom by reduction or by transfer of the chlorine to other amino compounds. However, between 1% and 16% of the amount of radioactivity detected in the blood was due to unreacted ^3H-N-chloropiperidine.

CONCLUSIONS

In this chapter a number of reactions of aqueous chlorine and chloramine with organic components of biological fluids in vivo and in vitro have been discussed. Many of them have been conducted using oxidant concentrations high enough so that products could be recognized in the highly complex organic mixtures of biological fluids, isolated, and identified. It is the goal of this research to extrapolate the effects of these reactions to oxidant concentrations used in toxicological studies and ultimately to oxidant concentrations ingested daily by humans.

Aqueous chlorine is an oxidizing agent as well as a chlorinating agent. Operationally, it has been convenient to identify the halogenated products of the reactions of chlorine disinfectants with organics, because the presence of a chlorine atom is a specific indicator that a chlorination reaction has taken place.[28] However, oxidation products that do not contain halogen atoms are also possible and may exhibit toxicological effects of their own. However, they will be far more difficult to identify.

ACKNOWLEDGMENT

This document has been reviewed in accord with U.S. Environmental Protection Agency policy through CR-813092 to Old Dominion University Research Foundation and approved for publication. Mention of trade names or commercial products does not constitute endorsement or recommendation for use.

REFERENCES

1. Bogoch, A., R. Wilson, S. Fishman, M. R. Kliman, G. E. Trueman, and G. I. Norton. "The Stomach and Duodenum," in *Gastroenterology,* A. Bogoch, Ed. (New York: McGraw-Hill Book Company, 1973), Chapter 11.
2. Forstner, G. G., M. A. Mullinger, and A. Bogoch. "Nutrition and Metabolism in Relation to Gastroenterology," in *Gastroenterology,* A. Bogoch, Ed. (New York: McGraw-Hill Book Company, 1973), Chapter 4.
3. Lehninger, A. L. *Biochemistry,* 2nd ed. (New York: Worth Publishers, Inc., 1977), 45.
4. Brooks, F. P. "Gastric Secretion: The Hypothalamus and the Vagus," in *The Stomach,* C. M. Thompson, D. Berkowitz, and E. Polish, Eds. (New York: Grune & Stratton, 1967), 119–123.
5. Fukayama, M. Y., H. Tan, W. B. Wheeler, and C. Wei. "Reactions of Aqueous Chlorine and Chlorine Dioxide with Model Food Compounds," *Environ. Health Pers.* 69:267–274 (1986).
6. Scully, F. E., Jr., K. Mazina, D. Sonenshine, and F. Kopfler."Quantitation and Identification of Organic N-Chloramines Formed in Stomach Fluid on Ingestion of Aqueous Hypochlorite," *Environ. Health Pers.* 69:259–265 (1986).
7. Clark, J. M., Jr., and R. L. Switzer. *Experimental Biochemistry,* 2nd ed. (San Francisco: W. H. Freeman and Co., 1977), 12.
8. Scully, F. E., Jr., and K. Mazina, unpublished results.
9. Morris, J. C. "Kinetics of Reactions Between Aqueous Chlorine and Nitrogen Compounds," in *Principles and Applications of Water Chemistry,* S. D. Faust and J. V. Hunter, Eds. (New York: John Wiley & Sons, 1967), 23–53.
10. Higuchi, T., A. Hussain, and I. H. Pitman. "Mechanism and Thermodynamics of Chlorine Transfer Among Organohalogenating Agents. Part IV. Chlorine Potentials and Rates of Exchange," *J. Chem. Soc. (B)* 1969:626–631 (1969).
11. Scully, F. E., Jr., J. P. Yang, K. Mazina, and F. B. Daniel. "Derivatization of Organic and Inorganic N-Chloramines for High-Performance Liquid Chromatographic Analysis of Chlorinated Water," *Environ. Sci. Technol.* 18:787–792 (1984).
12. Scully, F. E., Jr., K. Mazina, D. E. Sonenshine, and F. B. Daniel. "Reactions of Hypochlorite and Organic N-Chloramines in Stomach Fluid," in *Water Chlorination: Chemistry, Environmental Impact and Health Effects, Vol. 5,* R. L. Jolley, R. J. Bull, W. P. Davis, S. Katz, M. H. Roberts, Jr., and V. A. Jacobs, Eds. (Chelsea, MI: Lewis Publishers, Inc., 1985), 175–184.

13. Mink, F. L., W. E. Coleman, J. W. Munch, W. H. Kaylor, and H. P. Ringhand. "In Vivo Formation of Halogenated Reaction Products Following Peroral Sodium Hypochlorite," *Bull. Environ. Contam. Toxicol.* 30:394–399 (1983).

14. Trehy, M. L., and T. L. Bieber. "Detection, Identification, and Quantitative Analysis of Dihaloacetonitriles in Chlorinated Natural Waters," in *Advances in the Identification and Analysis of Organic Pollutants in Water, Vol. 2,* L. H. Keith, Ed. (Ann Arbor, MI: Ann Arbor Science Publishers, Inc., 1981), 941–975.

15. Bieber, T. I. and M. L. Trehy. "Dihaloacetonitriles in Chlorinated Natural Waters," in *Water Chlorination: Environmental Impact and Health Effects, Vol. 4,* R. L. Jolley, W. A. Brungs, J. A. Cotruvo, R. B. Cumming, J. S. Mattice, and V. A. Jacobs, Eds. (Ann Arbor, MI: Ann Arbor Science Publishers, Inc., 1983), 85–96.

16. Scully, F. E., Jr., R. Kravitz, C. Dean Howell, M. A. Speed, and R. P. Arber. "Contributions of Proteins to Formation of Trihalomethanes on Chlorination of Natural Waters," in *Water Chlorination: Chemistry, Environmental Impact and Health Effects, Vol. 5,* R. L. Jolley, R. J. Bull, W. P. Davis, S. Katz, M. H. Roberts, Jr., and V. A. Jacobs, Eds. (Chelsea, MI: Lewis Publishers, Inc., 1985), 807–820.

17. Morris, J. C., N. M. Ram, B. Baum, and E. Wajon. "Formation and Significance of N-Chloro Compounds in Water Supplies," EPA-600/2–80–031 (Cincinnati: U.S. Environmental Protection Agency, 1980).

18. Stevens, A. A., C. J. Slocum, D. R. Seeger, and G. G. Robeck. "Chlorination of Organics in Drinking Water," in *Water Chlorination: Environmental Impact and Health Effects, Vol. 1,* R. L. Jolley, Ed. (Ann Arbor, MI: Ann Arbor Science Publishers, Inc., 1978), 77–104.

19. Isaac, R. A., and J. C. Morris. "Transfer of Active Chlorine from Chloramine to Nitrogenous Organic Compounds. 1. Kinetics," *Environ. Sci. Technol.* 17:738–742 (1983).

20. Isaac, R. A., and J. C. Morris. "Transfer of Active Chlorine from Chloramine to Nitrogenous Organic Compounds. 2. Mechanism," *Environ. Sci. Technol.* 19:810–814 (1985).

21. Bercz, J. P., and R. Bawa. "Iodination of Nutrients in the Presence of Chlorine Based Disinfectants Used in Drinking Water Treatment," *Toxicol. Lett.* 34(2–3):141–147 (1986).

22. Hand, Vincent C., and D. W. Margerum. "Kinetics and Mechanisms of the Decomposition of Dichloramine in Aqueous Solution," *Inorg. Chem.* 22:1449–1456 (1983).

23. Mazina, K. E., F. E. Scully, Jr., and H. P. Ringhand. "Chlorinated Organic By-Products of the Reaction of N-Chloropiperidine with Stomach Fluid from Laboratory Animals," presented at the Sixth Conference on Water Chlorination: Environmental Impact and Health Effects, Oak Ridge, TN, May 3–8, 1987.

24. Soper, G., and G. F. Smith. "The Halogenation of Phenols," *J. Chem. Soc.* 1582 (1926).

25. Suh, D. H., and M. S. Abdel-Rahman. "Kinetics Study of Chloride in Rat," *J. Toxicol. Environ. Health* 12:467–473 (1983).

26. Abdel-Rahman, M. S., D. M. Waldron, and R. J. Bull. "A Comparative Kinetics Study of Monochloramine and Hypochlorous Acid in Rat," *J. Appl. Toxicol.* 3(4):175–179 (1983).
27. Anbar, M., S. Guttman, and R. Rein. "The Isotopic Exchange between Hypochlorite and Halide Ions. II. The Exchange Between Hypochlorous Acid and Chloride Ions." *J. Am. Chem. Soc.* 81:1816–1821 (1959).
28. National Academy of Sciences. *Drinking Water and Health, Vol. 7* (Washington: National Academy Press, 1987), Chapter 3.

SECTION V

Ozonation By-Products

CHAPTER 13

Ozone Oxidation Products—Implications for Drinking Water Treatment

Rip G. Rice, Rice International Consulting Enterprises, Ashton, Maryland

INTRODUCTION

Implementation of the regulatory requirements of the Safe Drinking Water Act Amendments of 1986 by the U.S. Environmental Protection Agency is encouraging U.S. water utilities to consider the incorporation of ozone into drinking water treatment processing, primarily for disinfection, but also for oxidation. One question frequently asked about ozone relates to the oxidation products caused by its use, *and* the toxicological significance of these oxidation products.

In this chapter, the major findings of a detailed literature survey conducted through 1985[1] will be reviewed briefly and the implications for water treatment with ozone will be discussed. In this detailed literature review, the authors surveyed oxidation products produced in water by ozone, chlorine, chlorine dioxide, chloramine, and potassium permanganate. Each of these disinfectants/oxidants can be applied at various points in a conventional water treatment process.

OZONATION CONDITIONS AND POINTS OF OZONATION

In operational drinking water treatment plants currently employing ozone for a variety of treatment purposes, normal ozonation conditions involve applied ozone dosages of 1 to 5 mg/L over contact times ranging from 5 to 15 minutes. These conditions generally are sufficient to convert easily oxidizable inorganic water constituents to their highest oxidation state quite rapidly. On the other hand, these ozonation conditions rarely are sufficient to oxidize all organic materials completely to carbon dioxide and water. Instead, most of the products of ozonation of organic materials are partially oxidized acids, aldehydes, ketones, and alcohols.

Figure 1 shows a schematic diagram of the conventional drinking water treatment process with points at which ozone is added, depending upon the purpose(s) for ozonation. For example, microflocculation and oxidation of iron, manganese, algae, and taste- and odor-causing organics with ozone generally is conducted in the raw water or in the rapid mix. Oxidation of more refractory organics (color, THM precursors, certain refractory organics such as some pesticides, herbicides, and insecticides, etc.) generally is conducted after sedimentation but prior to filtration. Primary disinfection with ozone normally is practiced after filtration, followed by detention (to allow the ozone residual to dissipate), then addition of small amounts of chlorine, chloramine, or chlorine dioxide to provide a secondary (residual) disinfectant in the distribution system.

OZONE DECOMPOSITION PRODUCTS

Oxidation products found in water after application of ozone include decomposition products of ozone itself and oxidation products from the organic materials present. Decomposition products from ozone itself include oxygen (O_2), the ozonide radical anion (O_3^-), superoxide ions (O_2^-), the perhydroxyl anion (HO_2^-), and the hydroxyl free radical (OH). In addition, hydrogen peroxide is produced in small quantities by decomposition of ozone in water, or as a by-product of ozone oxidation of dissolved organic materials.[2,3]

Most of these intermediates are highly reactive, particularly in water at ambient temperatures, and their rapid decomposition leads to oxygen or hydroxide ions as final, stable decomposition products in aqueous solutions. The hydroxyl free radical is unusually reactive, having an oxidation potential greater than that of ozone itself (2.80 vs 2.07 V). Consequently, deliberate formation of hydroxyl free radicals can assist in oxidizing organic materials.

Decomposition of ozone in water is initiated by hydroxide ion, ultraviolet radiation, or hydrogen peroxide to produce these intermediate species, particularly the hydroxyl free radical. On the other hand, the decomposition of ozone is affected by the presence of carbonate and bicarbonate ions. For example, bicarbonate ions interrupt the cyclic decomposition of ozone by reacting with the transient hydroxyl free radical:[3]

$$HCO_3^- + \bullet OH \rightarrow OH^- + \bullet HCO_3$$

The net effect is to retard the decomposition of ozone. Staehelin and Hoigné[2] have shown that increasing the concentration of carbonate ions at

POINT(S) OF OZONATION

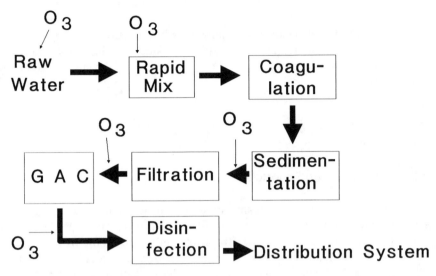

Figure 1. Conventional water treatment process showing points of ozonation. Granular activated carbon (GAC) adsorption is an optional addition to the conventional treatment process.

pH 8 doubled the half-life of ozone. Organic impurities present also scavenge hydroxyl free radicals, and themselves become oxidized in the process.

INORGANIC OXIDATION PRODUCTS OF OZONE WATER TREATMENT

As indicated earlier, ozone is capable of oxidizing inorganic ions in water rapidly to their highest oxidation states. Thus, ferrous ions are oxidized rapidly to ferric ions, which then hydrolyze and precipitate from solution. Manganous $(2+)$ ions are oxidized to the manganic $(4+)$ state, which hydrolyze to produce insoluble manganic oxides. However, further ozonation can continue the oxidation to the soluble permanganate stage $(7+)$. Nitrite ions are rapidly converted to nitrate ions by ozone, and sulfide ion is oxidized through elemental sulfur to sulfite, then to sulfate ions. Arsenious ions $(3+)$ are rapidly oxidized by ozone to the arsenate ions $(5+)$, in which stage they can be precipitated readily by polyvalent cations such as calcium.

Bromide ion is oxidized readily by ozone to produce hypobromite and then bromate ions:[4]

$$O_3 + OBr^- \rightarrow [O_2 + BrOO^-] \rightarrow 2O_2 + Br^- \ (77\%)$$

$$2O_3 + OBr^- \rightarrow 2O_2 + BrO_3^- \ (23\%)$$

Hypochlorite ion also oxidizes bromide ions by similar mechanisms. Formation of hypobromite ion and its hydrolysis product hypobromous acid are the source of brominated organic products formed during the chlorination of drinking water.

ORGANIC OXIDATION PRODUCTS OF OZONE WATER TREATMENT

Aliphatic Oxidation Products

Oxidation products from organic impurities in water usually are acids, ketones, aldehydes, and alcohols. So-called ultimate oxidation products of organic materials are CO_2, water (from total mineralization), oxalic acid, acetic acid, and hydrogen peroxide. However, ozonation conditions generally employed in treating drinking water rarely are sufficient to form high concentrations of these ultimate oxidation products.[5] Depending upon the amount of ozone added and the length of ozone contacting time (as well as the precursor materials present), the aldehydes produced upon ozonation can range in length from C-7 (heptanal) to C-15 (pentadecanal), down to formaldehyde, glyoxal, glyoxylic acid, and the like. Although to date, no unsaturated aldehydes have been isolated from ozonation experiments, some unsaturated aldehydes (enals) are known to be severe hepatotoxins.[3] These can be reactive with some biological molecules, blocking protein and DNA synthesis. Therefore, the search for such enals following ozonation should be continued.[6]

It is well-known that the ozonation of isolated unsaturated bonds

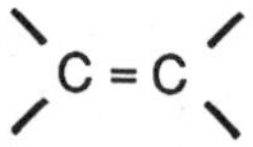

produces aldehydes or ketones, depending upon the carbon substituents:

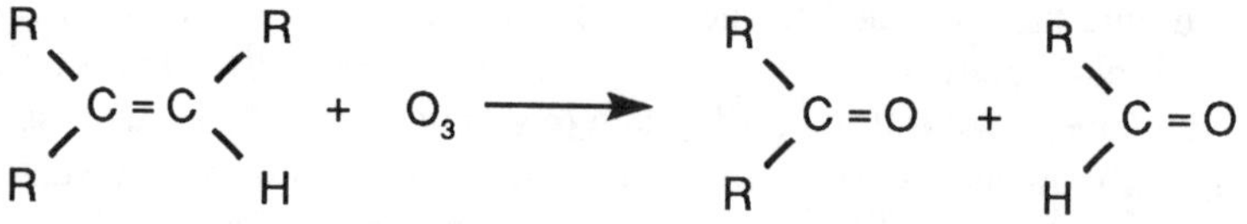

On the other hand, saturated aliphatic compounds are much less reactive with ozone, particularly under drinking water ozonation conditions of low applied ozone dosages and relatively short contact times. Some lower-molecular-weight saturated aliphatics (i.e., trihalomethanes such as chloroform; carbon tetrachloride; trichloroethane, etc.) which are not readily water soluble and are volatile, can be at least partially removed from aqueous solution during ozonation, although primarily by physically air-stripping, as opposed to ozone oxidation.

Aromatic Oxidation Products

Oxidation of aromatic compounds proceeds in stages, ring-hydroxylation occurring first, followed by quinone formation, then ring rupture to produce aliphatic, saturated, 2- and 4-carbon materials such as muconic acid, maleic acid, tartaric acid, glyoxal, glyoxylic acid, fumaric acid, ketomalonic acid, muconaldehyde, malealdehyde, oxalic acid, formic acid, CO_2, and hydrogen peroxide. Intermediate aromatic compounds which have been isolated from the ozonation of benzene and/or phenol include catechol, resorcinol, hydroquinone, and o- and p-benzoquinones.[1]

It is important to realize that the rates of initial oxidation of aromatic compounds may be rapid, but that continued oxidation with ozone proceeds at slower rates. For example, ozonation of phenol itself to produce hydroxylated aromatic intermediate compounds occurs at k values above 1,000 $M^{-1}s^{-1}$, but formation of CO_2 and water from the intermediate products proceeds at rates having k values of less than 10 $M^{-1}s^{-1}$.[7]

Ozonation of aromatic compounds using very small levels of oxidant also can result in oxidative coupling to produce polymeric compounds. Duguet et al.[8] have shown this type of reaction to occur with 2,4-dichlorophenol at elevated pH values. These oxidatively coupled materials have decreased solubilities in water over those of the starting compounds, and can be readily filtered from the ozonized solution.

In general, the order of reactivity of organic species to ozone oxidation is as follows:

$$alkenes > aromatics >>> alkanes$$

Fronk[9] recently has published an excellent compendium of the relative reactivities of a number of volatile organic compounds with ozone in water as determined under drinking water ozonation conditions.

Ozone-Refractory Organics

Recently, the U.S. Environmental Protection Agency[10] has listed some pesticides as Synthetic Organic Chemicals (SOCs) which will be the subject of regulatory actions in the near future. Many of these pesticidal compounds are reactive to ozone, at variable rates; however, some are unreactive, and the reactivity of toxaphene to ozone is unknown as yet. Those compounds known to be reactive with ozone are: lindane[11] (fast at pH 9 and 11; very slow at pH 4), chlorobenzene[12] (producing the same oxidation products as phenol, plus chloride ion and chlorophenols), o-dichlorobenzene[13] (95% destroyed in 15 min), pentachlorophenol[14] (very fast), ethylbenzene[15] (fast), toluene[15] (fast), styrene[16] (producing benzaldehyde and benzoic acid), xylene[15] (fast), cis- and trans-1,2-dichloroethylene[15] (fast), acrylamide and methoxychlor (C. A. Fronk, U.S. Environmental Protection Agency, Cincinnati, OH, 1987, private communication), and heptachlor.[17]

SOCs known to be unreactive (or only very slowly reactive) with ozone are: chlordane, PCBs, heptachlorepoxide, dibromochloropropane, 1,2-dichloropropane, epichlorohydrin, and ethylene dibromide.[18] Heptachlor is a listed SOC which is quite reactive with ozone, and is readily "removed or destroyed" by oxidation.[17] However, its primary oxidation product is heptachlorepoxide (also listed), obtained in nearly quantitative yield, which is relatively stable to further ozonation. This epoxide also is more toxic than is the original heptachlor.[17]

In laboratory studies, other epoxides have been isolated and identified, usually in trace amounts, from ozonation of the following compounds: oleic acid,[19] abietic acid,[19] cholesterol,[19] and acenaphthene.[20] Except for heptachlor, epoxides produced usually are formed in trace quantities in water, by ozone, but also by chlorine. With oleic acid, the same epoxide has been isolated in trace amounts during reactions with ozone, chlorine, chlorine dioxide, and chloramine.[19]

The pesticides parathion and malathion initially produce the corresponding oxons (paraoxon and malaoxon) upon ozonation. These are at least as toxic as the starting thions. Upon continued ozonation the intermediate oxons form phosphoric acid as one of the final oxidation products.[21] Other intermediates isolated from the ozonation of parathion include 2,4-dinitrophenol, picric acid, and sulfuric acid.[22]

Thus, it is important for the water treatment specialist to be aware of as many of the specific chemical compounds which may be present in his raw water as possible, as well as their oxidation chemistries, in order to be able to cope with pollutants without creating other pollution problems. It is also important to recognize that ozone is not alone in its ability to produce oxidized materials which have toxicological significance. Other strong oxidizing/disinfecting agents, including chlorine, produce similar, nonchlorinated oxidation products.

IMPLICATIONS FOR WATER TREATMENT

Formation of Aldehydes

One very significant implication of the production of aldehydes during ozonation (or chlorination) of drinking water is the ability of aldehydes to form nitriles upon subsequent treatment with chlorine and ammonia (chloramination).[23,24] For example:

$$0°C$$
$$RCH=O + NH_2Cl \rightarrow R-CH=N-Cl \text{ (a chloroimine)}.$$

The corresponding nitrile can be produced by loss of HCl from the chloroimine intermediate. Nitriles can be expected to impart some toxicological effect to the water by virtue of the presence of the nitrile grouping itself. In addition, nitriles can hydrolyze in aqueous solution, to produce the corresponding acids. If the nitriles are chlorinated, then chlorinated aliphatic acids can be produced by this mechanism.

A recent EPA announcement[25] lists several halogenated nitriles and halogenated aliphatic acids as having been found in drinking water supplies. EPA is proposing to set MCLs (Maximum Contaminant Levels) for the following discrete compounds in these classes which have been identified as being present in some chlorinated drinking water supplies:

- mono-, di-, and trichloroacetic acids
- trichloroacetaldehyde (chloral hydrate)
- bromochloroacetonitrile
- dichloro- and dibromoacetonitriles

Another important aspect of the formation of aldehydes is that they generally are more biodegradable than their precursors. This means that if the produced aldehydes are not removed from solution, they can give rise to bioregrowths in drinking water distribution systems. Such bioregrowths can be prevented by adding a bactericidal (chlorine, chlorine dioxide) or bacteriostatic (chloramine) agent to the treated water. In the case of chlorine, however, which is currently the most frequently used disinfectant in U.S. water treatment systems, the possibility exists that many of the aldehydes produced upon ozonation also are THM precursors.

In fact, several early studies of ozone oxidation followed immediately by chlorination showed *increased* levels of THMs rather than decreased levels.[26] Increased THM levels might be attributed, at least partially, to the formation of aldehydes by ozonation. Another possibility is hydroxylation of aromatic compounds to produce m-dihydroxy aromatic derivatives, which are known THM precursors.[27]

Although the aldehydes produced contain polar carbonyl groupings, they are nevertheless not easily removed during the flocculation step by complexation with aluminum or iron salts.[28] A convenient and more appropriate method for removal of the formed aldehydes is by incorporating a biological treatment step in the water treatment process following ozone oxidation. If ozonation is employed in a conventional treatment process before rapid sand or dual media filtration, usually a considerable amount of biodegradation of partially oxidized organic constituents will occur in the filter media, provided that no residual disinfectant (chlorine) is present.

As a recent case in point, Lykins et al.[28] conducted a one-year pilot plant study of the effects of four disinfectants (ozone, chlorine, chlorine dioxide, and monochloramine) on drinking water at the Jefferson Parish (LA) water treatment plant. The four disinfectants were applied at various dosages over 30 minutes' contact time prior to filtration through sand and then GAC (granular activated carbon). Toxicological effects were studied in each of the disinfected water streams, and the major organic oxidation products were isolated and identified in the disinfected water streams at various points after application of the disinfectants/oxidants.

As anticipated, ozone provided the best disinfection performance. Additionally, TOC (total organic carbon) concentration was reduced 38%, and the concentrations of selected nonvolatile organics (atrazine, alachlor, total chlorinated hydrocarbon insecticides, total alkylbenzenes, total alkanes, total phthalates, total chlorobenzenes, and total nitrobenzenes) also were lowered by 11% to 84% after ozonation (compared with *increases* of 43% to 100% of total alkylbenzenes, total chlorobenzenes, total nitrobenzenes, and total alkylaldehydes upon chlorination). No mutagenicity was observed in the ozonated water. It should be borne in mind, however, that 30 minutes of ozone contact time is much longer than normally practiced in ozone treatment of drinking water. Such a long contact time is expected to produce considerable oxidation of the organic materials present.

On the other hand, the concentrations of aldehydes (octanal plus nonanal) increased by 144% upon ozonation. In the chlorinated stream, the concentrations of these two nonchlorinated aldehydes increased by 56%. This indicates that aldehyde formation, although greater with ozone, is not unique to ozonation, but is caused with chlorination as well.

Passage of the ozonized water samples through a rapid sand filter reduced the concentration of these aldehydes by 62%. Chlorinated samples experienced a 26% reduction in aldehyde concentrations under the same conditions. These reductions in aldehyde concentrations are attributed to biological activity in the sand filters.[28]

If GAC filtration follows sand filtration, ozone oxidation can be expected to promote even more biological activity in the GAC filter, because better housing is provided for microorganisms on GAC particles than on sand. Thus, the biological conversion of oxidized water impurities to CO_2 and water will be greater during passage through GAC media. Similar aldehyde-removal effects have been noted by Van Hoof et al.[29] in the Netherlands. During ozonation of various raw waters, these investigators formed C1-C3 aldehydes and C1-C3 acids. The concentrations of these organic oxidation products were found to increase as the applied ozone dosage increased from 1 mg/L to 6.7 mg/L. During subsequent treatment of these ozonized waters, the concentrations of aldehydes and acids were reduced in the dual media filters, in the water storage tank following filtration, and in the GAC filters following storage. Removal of aldehydes and acids is ascribed to biological conversion to CO_2 and water.

FORMATION OF MUTAGENICITY UPON OZONATION

Studies by Van Hoof et al.[30] have shown that waters ozonized with low applied dosages of ozone (1 mg/L) produce waters having relatively high levels of mutagenicity (by the Ames test), whereas waters treated with higher applied ozone dosages (4 mg/L) contain much lower levels of mutagenicity. Kool et al.[31] extended and confirmed these observations. With chlorine, these authors found large increases in both mutagenicity and TOCl (total organic chlorine) concentrations. Chlorine dioxide applied at dosages of 5–15 mg/L produced slight increases in TOCl concentrations, but large increases in mutagenicity. When applied at lower levels, i.e., < 5 mg/L, chlorine dioxide produced only slight increases in TOCl concentrations, and slight increases in, slight decreases in, or no effect on mutagenicity.

Ozone applied at the rate of 3 mg/L produced no increase in TOCl concentration, and caused a slight increase in mutagenicity. However, when larger dosages of ozone were applied (i.e., 10 mg/L), mutagenicity was reduced or completely removed; at the same time, no increase in TOCl concentration was observed.

French workers also have observed similar trends of increased mutagenicity of waters treated with small amounts of ozone versus no increase or decreases in mutagenicity of waters treated with larger dosages of ozone. Bourbigot et al.[32] developed the data shown in Table 1 during treatment of surface waters with ozonation followed by GAC adsorption. This indicates that GAC adsorption placed after ozonation removes all traces of mutagenicity which ozone (or other oxidants) may produce.

Cognet et al.[33] showed that in waters containing 5–15 mg/L of dissolved organics, treatment with ozone or with chlorine produced increases in mutagenicity of the treated water. However, passage of these ozone- or chlorine-treated waters through GAC filters removed the mutagenic activity formed

Table 1.   Mutagenicity Produced by Ozonation and Removed by GAC[32]

Ozone Dosage Applied (mg/L)	Mutagenicity	
	After Ozonation	After GAC
0.75	slight increase	no activity
1.5	slight increase	no activity
3.0	no activity	no activity

by either strong disinfectant. These researchers[33] concluded, along with the Dutch,[31] that incorporating GAC filtration after ozonation not only provides an efficient biological treatment step to remove partially oxidized organic materials from the water and reduce the potentials for bioregrowths in the distribution system, but also removes any mutagenic activity present.

WATER TREATMENT PROCESSING CONSIDERATIONS

Ozone is used for a variety of purposes in water treatment.[34,35] Raw waters are treated with ozone before or during the rapid mix for microflocculation, oxidation of iron and manganese, turbidity control, and so on. Before filtration in a conventional treatment process, ozone is applied for oxidation of organic contaminants, color removal, taste and odor control, and so on. After filtration, ozone generally is applied for disinfection. However, the dosages of ozone applied and the contact times designed into the treatment process for ozonation at these various stages are quite different. Not paying attention to the distinctions between the various ozonation conditions can lead to apparent failure of ozone to accomplish one or more of its intended purposes.

Ozone For Microflocculation

When ozone is used to promote microflocculation of humic substances early in the water treatment process, the ozonation objective is only partial oxidation of the dissolved organics, without reducing their molecular weights.[36] Addition of polar carbonyl, carboxyl, or hydroxy moieties to the organic structure upon ozonation makes the large organics more easily flocculated upon addition of aluminum or iron flocculants. Excessive ozonation at this stage of treatment provides extensive oxidation of the organic components, producing lower-molecular-weight fragments which defeat the purpose of flocculation and are not easily removed during the flocculation step.

An additional advantage of low-level ozonation if phenolics or humic materials are present is oxidative coupling.[8] For example, when 2, 4-dichlorophenol was ozonized at low dosage levels in water, approximately 8% of the initial 2,4-chlorophenol was converted to polymerized material. About 18% of this material has apparent molecular weight above 5000 daltons.

Raw waters contain the highest concentrations of organic materials. Application of very low ozone dosages at this point produces the maximum amount of mutagenicity, the maximum amount of oxidative coupling (to yield higher-molecular-weight products more readily removed through flocculation/filtration), the maximum amount of flocculatable materials, and the minimum amount of discrete organic oxidation by-products.

Ozone For Prefiltration Oxidation

After flocculation and sedimentation, but prior to filtration, ozone can be used to oxidize some of the more refractory organic materials which may be present. Under these circumstances, with organics oxidation as the primary objective, higher applied ozone dosages will be required than for the earlier microflocculation application. Under these circumstances, mutagenic activity either will not be produced, or will be produced, then immediately destroyed. In addition, low-molecular-weight by-products (i.e., aldehydes and acids) will be produced which are biodegradable, but are not easily flocculated.

Ozone For Postfiltration Disinfection

If prefiltration ozone oxidation is not necessary, postfiltration primary disinfection will be required. Ozone has been used for many years for this purpose, normally applied in two adjacent ozone contact chambers.[37] Water entering the first contact chamber has the higher ozone demand. In this first chamber, the short-term ozone demand of the water is satisfied, and a specified dissolved ozone concentration (usually 0.4 mg/L) is developed over a period of 4–5 minutes. In the second contact chamber this residual ozone concentration of 0.4 mg/L (or higher) is maintained for a minimum of four minutes, and more normally over 6–10 minutes.

These "viral inactivation conditions" were developed by French public health scientists during the mid-1960s[38,39] to guarantee 99.9% inactivation of poliovirus types I, II, and III. These "ozone disinfection conditions" subsequently have been specified in all French drinking water treatment plants using ozone for disinfection, and they have been adopted as guidelines by the World Health Organization as well.

At this point in the water treatment process (water has been flocculated, settled, and filtered), the water contains the least amount of organic materials. Consequently, its ozone demand will be the lowest, and the production of organic oxidation products will be minimal. However, some of those organics which remain may be oxidized by ozone, at least partially.

If the previous treatment steps have not removed most of the organic materials effectively, then the concentration of oxidizable organics will be higher at the point of postfiltration ozone disinfection. In this situation, ozone disinfection also will produce considerable oxidation of residual organics, perhaps to increase the THMFP (trihalomethane formation potential) or TOXFP (total organic halogen formation potential), but certainly to produce aldehydes, acids, and ketones as organic oxidation products. If GAC adsorption is required to remove specific organic compounds, disinfection also will be required after the GAC filter/adsorbers. This is to destroy microorganisms which contaminate the GAC effluent.

The objective of ozone disinfection at this point in the treatment process is primary disinfection of microorganisms. Although in the past, disinfection has meant simply the killing of coliform bacteria, newly proposed EPA regulations[40] redefine disinfection to mean removal or inactivation of *Giardia* cysts and enteric viruses, in addition to the destruction of coliform bacteria, *Legionella* bacteria, and heterotrophic plate count organisms.

EPA has proposed that rather than monitoring for all of these organisms, a water utility practicing disinfection to meet the new EPA drinking water disinfection requirements will only have to monitor and report data showing the concentration of disinfectant and the contact time of the disinfectant at its appropriate pH range. In this manner, a "CT value" for each disinfectant (ozone, chlorine, chlorine dioxide, chloramine) can be calculated readily. EPA's new surface water treatment regulations[40] require the inactivation/removal of 99.9% of *Giardia* cysts and 99.99% of enteric viruses in addition to the destruction of bacterial organisms. Specific CT values for each disinfectant applied in a specific pH range and at specific water temperatures have been developed by EPA so that water utilities may design their disinfection systems appropriately and provide monitoring data to prove they are in compliance with the new disinfection requirements.

With ozone, for example, CT values range from 4.5 mg/L-min at 0.5°C to 1 mg/L-min at 25°C to provide 99.9% inactivation of *Giardia* cysts and (simultaneously) 99.99% inactivation of enteric viruses. When considering ozone for primary disinfection (pre-or postfiltration), it will be important for the water treatment system designer to provide for a dissolved ozone concentration of, say, 0.5 mg/L for at least 9 minutes at 0.5°C to guarantee a CT value of 4.5 mg/L-min, or to provide a contact time of at least 2 minutes at 25°C to provide a CT value of 1 mg/L-min at this temperature. In turn, this will require monitoring dissolved ozone levels at two points in the ozonation system, in order to be assured that the required dissolved ozone concentration is maintained for the required contact time.

Because of the very low CT values for ozone which provide the required degree of disinfection of *Giardia* cysts and/or enteric viruses, it is possible to attain primary disinfection during prefiltration ozone oxidation for some purpose other than disinfection.

In any event, water exiting the ozone contact chamber will contain some quantity of residual ozone. Following ozone disinfection, a secondary (residual) disinfectant is required by EPA for protection of the water distribution system. The secondary disinfectant may be chlorine, chlorine dioxide, or monochloramine. However, because of the reactivity of ozone with hypochlorite ion, chlorine dioxide, and chloramine,[4] it will be important to be certain that residual dissolved ozone has dissipated from solution after postfiltration ozone disinfection, prior to adding the secondary disinfectant. The absence of residual ozone at this point can be assured by a number of procedures: physically degassing the ozonized water, allowing the excess ozone to decompose in the clearwell prior to adding the chlorinous secondary disinfectant, or by incorporating a GAC filter/adsorber in the treatment system.

Benefits of GAC Adsorption Following Ozone Treatment

Not only will a GAC filter positioned after ozone oxidation or disinfection destroy excess dissolved ozone, but it will also adsorb refractory organic materials, and provide a biological treatment step for further converting partially oxidized organic materials to CO_2 and water microbiologically. In point of fact, the incorporation of GAC filtration following ozone oxidation provides a type of insurance policy to maximize the benefits of ozonation (removal of oxidation products, removal of any mutagenicity produced by low level ozonation, avoidance of ozone destruction of secondary disinfectants, destruction of excess ozone, etc.).

Dutch water technologists,[30-32,41] among others, have shown that ozonation maximizes the production of Assimilable Organic Carbon (AOC), but that passing ozonized water through GAC filters removes the AOC biologically. The GAC effluents now exert minimal chlorine demand to produce a stable residual in the distribution system. In addition, because the AOC produced by ozonation has been removed biologically in the water treatment plant, there is minimal bioregrowth in the distribution systems. In fact, the Dutch prefer not to add a residual disinfectant to water treated by ozone/GAC, to avoid producing additional AOC.

REGULATORY CONSIDERATIONS

EPA's new primary disinfection requirements[40] can be met by ozone while ozone is being used simultaneously as a prefiltration oxidizing agent. This is because of the low CT values of ozone (4.5 to 1 mg/L-min) to attain the required levels of *Giardia,* virus, and bacterial organism removal/inactivation. For ozone oxidation, it is customary to apply ozone for contact times of 10–15 minutes. During this time, dissolved ozone levels of 0.5 to 1.0 mg/L can be obtained routinely. Thus CT values of 5 to 15 mg/L-min are obtained routinely during ozone oxidation. It remains for the water utility simply to monitor the dissolved ozone residual at two appropriate points in the ozonation system to make the CT calculation, and provide the required CT data to regulatory monitoring agencies.

On the other hand, there are two preozonation applications under which primary disinfection with ozone cannot be combined effectively: these are the circumstances in which ozone is used for microflocculation or for oxidation of iron and manganese. When ozone is used for microflocculation, the objective is to add only trace quantities of ozone to produce partial oxidation of the dissolved organics. It is important to avoid overozonation of the organics, thus lowering their molecular weights. Under such conditions, a measurable residual of dissolved ozone rarely is established; even if residual ozone can be measured, its concentration should be well below 0.1 mg/L. Such a low level of residual ozone would require contact times of 45 minutes to 10 minutes to attain CT values of 4.5 to 1 mg/L-min required to produce 99.9% inactivation of *Giardia* cysts and 99.99% inactivation of enteric viruses.

When ozone is employed to oxidize iron and manganese, insoluble products are produced which will foul residual ozone analytical probes. In this circumstance, even though an adequate residual of ozone may be present to provide the appropriate CT value, it is impossible to measure that residual with sufficient confidence to assure regulatory compliance.

POSITION OF NATIONAL ACADEMY OF SCIENCES REGARDING OZONE FOR TREATING DRINKING WATER

In a recent analysis of the current status of knowledge concerning the oxidation products from all water disinfectants/oxidants, the National Academy of Sciences[6] states that the nonchlorinated oxidation products from the use of chlorine, chlorine dioxide, and ozone isolated and tested to date "are similar to natural degradation products, and that water utilities should not allow the current lack of complete knowledge of ozone oxidation products to retard consideration of ozone as a water oxidant and/or disinfectant."

SUMMARY AND CONCLUSIONS

1. Ozone produces organic oxidation products which are nonhalogenated and which generally are biodegradable. Aliphatic aldehydes are a large proportion of the oxidation products, even from the decomposition of aromatic starting materials.

2. Although the aldehydes formed upon ozonation are not readily removed from solution by flocculation/filtration, they are readily decomposed biologically to CO_2 and water in sand filters (provided no residual disinfectant is present) or GAC filter/adsorbers.

3. If bromide ion is present in the water being ozonized, brominated organic materials can be produced, as a result of oxidation of bromide ion to hypobromite ion. Chlorine and chlorine dioxide also are known to oxidize bromide ion similarly. Once produced, halogenated organics are difficult to remove.

4. Some pesticidal compounds produce intermediate oxidation products upon ozonation which are more toxic than the starting pesticides. In the case of heptachlor, a nearly quantitative yield of heptachlorepoxide is produced. This epoxide is relatively stable to ozonation. Consequently, if heptachlor is present in the raw water, either a treatment technology other than ozonation must be employed, or removal of heptachlorepoxide must be provided.

5. Ozonation of the insecticides parathion and malathion produces paraoxon and malaoxon, respectively, which are at least as toxic as the initial thions. Continued ozonation, however, oxidizes these intermediates further, producing innocuous ultimate oxidation products.

6. Low-level ozonation (i.e., 0.5–1.5 mg/L applied ozone dosages) can produce low levels of mutagenicity. Higher applied ozone dosages (i.e., 3 mg/L) produce waters containing no mutagenicity. Following ozonation with GAC, adsorption/filtration also will eliminate the low level of mutagenicity produced by low-level ozonation.

7. In addition, GAC will adsorb organics which are not easily oxidized by ozone.

8. Because of its reactivity with hypochlorite anions, with chlorine dioxide, and with chloramines, ozone should be added to water in the absence of these chlorinous reagents. Otherwise, both ozone and the chlorine-containing material will be mutually destroyed.

9. Combining prefiltration ozone oxidation with primary disinfection in the same treatment step will eliminate the need for postfiltration primary disinfection with ozone, chlorine, or chlorine dioxide. Consequently, only a secondary disinfectant will be required. In the case of chlorine, this will result in lower concentrations of THMs and other halogenated organics, soon to be regulated by the U.S. Environmental Protection Agency.

REFERENCES

1. Rice, R. G., and M. Gomez-Taylor. "Occurrence of By-Products of Strong Oxidants Reacting with Drinking Water Contaminants—Scope of the Problem," *Environ. Health Pers.* 69:31–44 (1986).
2. Staehelin, J., and J. Hoigné. "Decomposition of Ozone in Water: Rate of Initiation by Hydroxide Ions and Hydrogen Peroxide," *Environ. Sci. Technol.* 16:676–681 (1982).
3. Glaze, W. H. "Drinking Water Treatment With Ozone," *Environ. Sci. Technol.* 21(3):224–230 (1987).

4. Haag, W. R., and J. Hoigné. "Kinetics and Products of the Reactions of Ozone with Various Forms of Chlorine and Bromine in Water," *Ozone Sci. Eng.* 6(2):103–114 (1984).

5. Gilbert, E. "Chemical Changes and Reaction Products in the Ozonization of Organic Water Constituents," in *Oxidation Techniques in Drinking Water Treatment*, W. Kühn and H. Sontheimer, Eds. U.S. EPA Report 570/9–79–020 (Washington, DC: U.S. Environmental Protection Agency, Office of Drinking Water, 1978), 232–270.

6. *Drinking Water and Health, Disinfectants and Disinfectant By-Products* (Washington, DC: National Academy Press, 1987).

7. Hoigné, J. "Mechanisms, Rates and Selectivities of Oxidations of Organic Compounds Initiated by Ozonation of Water," in *Handbook of Ozone Technology and Applications*, R. G. Rice and A. Netzer, Eds. (Ann Arbor, MI: Ann Arbor Science Publishers, Inc., 1982), 341–379.

8. Duguet, J. P., A. Bruchet, B. Dussert, and J. Mallevialle. "Formation of Aromatic Polymers During the Ozonation or Enzymatic Oxidation of Waters Containing Phenolic Compounds," this book, Chapter 14.

9. Fronk, C.A. "Destruction of Volatile Organic Contaminants in Drinking Water by Ozone Treatment," *Ozone Sci. Eng.* 9(3):265–287 (1987).

10. U.S. Environmental Protection Agency. "National Primary Drinking Water Regulations; Synthetic Organic Chemicals, Inorganic Chemicals and Microorganisms; Proposed Rule," *Federal Register* 50(219):46935–47022, Nov. 13, 1985.

11. Ishizaki, K., R. A. Dobbs, and J. M. Cohen. "Ozonation of Hazardous and Toxic Organic Compounds in Aqueous Solution," in *Ozone/Chlorine Dioxide Oxidation Products of Organic Materials*, R. G. Rice and J. A. Cotruvo, Eds. (Norwalk, CT: International Ozone Association, 1978), 210–226.

12. Bauch, H., B. Burchard, and H. M. Arsovic. "Ozone as an Oxidative Decomposition Material for Phenols in Aqueous Solutions," *Gesundheits-Ingenieur* 91(9):258 (1970).

13. Prengle, H. W., Jr., C. E. Mauk, and J. E. Payne. "Ozone/UV Oxidation of Chlorinated Compounds in Water," in *Forum on Ozone Disinfection*, E. G. Fochtman, R. G. Rice, and M. E. Browning, Eds. (Norwalk, CT: International Ozone Association, 1977), 286–295.

14. Schwartz, M., R. G. Rice, and A. Levy. "The Fate of Pentachlorophenol in a Biological Activated Carbon System," in *Sixth Ozone World Congress Proceedings*, (Norwalk, CT: International Ozone Association, 1983), 111–112.

15. Fronk, C. A. "Destruction of Volatile Organic Contaminants in Drinking Water by Ozone Treatment," *Ozone Sci. Eng.* 9(3):265–287 (1987).

16. Yocum, F. H. "Oxidation of Styrene with Ozone in Aqueous Solution," in *Ozone/Chlorine Dioxide Oxidation Products of Organic Materials*, R. G. Rice and J. A. Cotruvo, Eds. (Norwalk, CT: International Ozone Association, 1978), 243–263.

17. Hoffmann, J., and D. Eichelsdörfer. "On the Action of Ozone on Pesticides with Chlorinated Hydrocarbon Groups in Water," *Vom Wasser* 38:197–206 (1971).

18. Rice, R. G. "Identification and Toxicology of Oxidation Products Produced During Treatment of Drinking Water," Draft Report submitted to U.S. Environ-

mental Protection Agency, Office of Drinking Water, under Purchase Order No. 4W-2943 NTSX (1986).

19. Carlson, R. M., and R. Caple. "Chemical/Biological Implications of Using Chlorine and Ozone for Disinfection," U.S. EPA Report No. 600/3–77–066, June 1977. NTIS Report No. PB-270–694.

20. Chen, P. N., G. A. Junk, and H. J. Svec. "Reactions of Organic Pollutants. I. Ozonation of Acenaphthylene and Acenaphthene," *Environ. Sci. Technol.* 13:451–454 (1979).

21. Laplanche, A., G. Martin, and F. Tonnard. "Ozonation Schemes of Organophosphorus Pesticides. Application in Drinking Water Treatment," *Ozone Sci. Eng.* 6(3):207–219 (1984).

22. Richard, Y. and L. Brener. "Organic Materials Produced upon Ozonization of Water," in *Ozone/Chlorine Dioxide Oxidation Products of Organic Materials,* R. G. Rice and J. A. Cotruvo, Eds. (Norwalk, CT: International Ozone Association, 1978), 169–188.

23. Crochet, R. A., and P. Kovacic. "Conversion of o-Hydroxyaldehydes and Ketones into o-Hydroxyanilides by Monochloramine," *J. Chem. Soc. Chem. Commun.* (1973):716–717 (1973).

24. Hauser, C. R. and M. L. Hauser. "Researches on Chloramines. I. Orthochlorobenzal-chlorimine and Anisalchlorimine," *J. Am. Chem. Soc.* 52:2050–2054 (1930).

25. U.S. Environmental Protection Agency. "Drinking Water; Proposed Substitution of Contaminants and Proposed List of Additional Substances Which May Require Regulation Under the Safe Drinking Water Act," *Federal Register* 52(130):25719–25734, July 8, 1987.

26. Trussell, R. R,. and M. D. Umphres. "The Formation of Trihalomethanes," *J. Am. Water Works Assoc.* 70:604–612 (1978).

27. Rook, J. J. "Possible Pathways for the Formation of Chlorinated Degradation Products During Chlorination of Humic Acids and Resorcinol," in *Water Chlorination: Environmental Impact and Health Effects, Vol. 3,* R. L. Jolley, W. A. Brungs, and R. B. Cumming, Eds. (Ann Arbor, MI: Ann Arbor Science Publishers, Inc., 1980), 85–98.

28. Lykins, B. W., W. Koffskey, and R. G. Miller. "Chemical Products and Toxicologic Effects of Disinfection," *J. Am. Water Works Assoc.* 78(11):66–75 (1986).

29. Van Hoof, F., J. G. Janssens, and H. Van Dyck. "Formation of Oxidation By-Products in Surface Water Preozonation and Their Behaviour in Water Treatment," *Water Supply* 4:93–102 (1986).

30. Van Hoof, F., J. G. Janssens, and H. Van Dyck. "Formation of Mutagenic Activity During Surface Water Preozonation and Its Removal in Drinking Water Treatment," *Chemosphere* 14(5):501–510 (1985).

31. Kool, H. J., C. F. van Kreijl, and J. Hrubec. "Mutagenic and Carcinogenic Properties of Drinking Water," in *Water Chlorination: Chemistry, Environmental Impact and Health Effects, Vol. 5,* R. L. Jolley, R. J. Bull, W. P. Davis, S. Katz, M. H. Roberts, Jr., and V. A. Jacobs, Eds. (Chelsea, MI: Lewis Publishers, Inc., 1985), 187–205.

32. Bourbigot, M. M., M. C. Hascoet, Y. Levi, F. Erb, and N. Pommery. "Role of Ozone and Granular Activated Carbon in the Removal of Mutagenic Compounds," *Environ. Health Pers.* 69:159–163 (1986).

33. Cognet, L., Y. Courtois, and J. Mallevialle. "Mutagenic Activity of Disinfection By-Products," *Environ. Health Pers.* 69:165–175 (1986).

34. Rice, R. G., C. M. Robson, G. W. Miller, and A. G. Hill. "Uses of Ozone in Drinking Water Treatment," *J. Am. Water Works Assoc.* 73(1):44–57 (1981).

35. Rice, R. G. "Ozone in Practice: European and U.S. Experiences," in *AWWA Seminar Proceedings. Ozonation: Recent Advanced and Research Needs* (Denver, CO: American Water Works Association, 1986), 147–173.

36. Maier, D. "Microflocculation by Ozone," in *Handbook of Ozone Technology and Applications. Vol. II. Ozone for Drinking Water Treatment,* R. G. Rice and A. Netzer, Eds. (Stoneham, MA: Butterworth Publishers, 1984), 123–140.

37. Miller, G. W., R. G. Rice, C. M. Robson, R. L. Scullin, W. Kühn, and H. Wolf. "An Assessment of Ozone and Chlorine Dioxide Technologies for Treatment of Municipal Water Supplies," U.S. EPA Report No. EPA-600/2–78–147 (Cincinnati, OH: U.S. EPA, Water Engineering Research Lab., 1978).

38. Coin, L., C. Hannoun, and C. Gomella. "Inactivation of Poliomyelitis Virus by Ozone in the Presence of Water," *La Presse Médicale* 72(37):2153–2156 (1964).

39. Coin, L., C. Gomella, C. Hannoun, and J. C. Trimoreau. "Ozone Inactivation of Poliomyelitis Virus in Water," *La Presse Médicale* 75(38):1883–1884 (1967).

40. U.S. Environmental Protection Agency. "National Primary Drinking Water Regulations; Filtration and Disinfection; Turbidity, *Giardia lamblia,* Viruses, *Legionella,* and Heterotrophic Bacteria; Proposed Rule," *Federal Register* 52(212):42177–42222 (November 3, 1987).

41. Van der Kooij, D. "The Effect of Treatment on Assimilable Organic Carbon in Drinking Water," in *Treatment of Drinking Water for Organic Contaminants,* P. M. Huck and P. Toft, Eds. (Oxford, England: Pergamon Press, 1987), 317–328.

CHAPTER 14

Formation of Aromatic Polymers During the Ozonation of Enzymatic Oxidation of Waters Containing Phenolic Compounds

Jean-Pierre Duguet, Auguste Bruchet, Bertrand Dussert, and Joël Mallevialle, Centre de Recherche, Lyonnaise des Eaux-Degrémont, Le Pecq, France

INTRODUCTION

Natural or synthetic organic compounds are usually present in raw waters at low concentrations (ng/L to µg/L). Even at low concentrations these compounds may cause taste and odor problems or represent a health risk, based on toxic and mutagenic considerations. These compounds are not entirely removed on conventional treatment of potable water. One means of increasing the removal efficiency is to use new processes based on the polymerization of these compounds.

Recent progress in biotechnology has permitted the study of enzyme reactions on aromatic pollutants. Klibanov et al.[1-3] studied the removal of phenols and aromatic amines by horseradish peroxidase-hydrogen peroxide (HRP-H_2O_2). High concentrations of aromatics were removed in the form of non-soluble polymers. The polymerization also reduces compounds that are not substrates of this enzyme, but which polymerize with radicals produced by enzymatic reaction.

Maloney et al.[4] showed that enzymatic reaction using peroxidase is efficient in eliminating traces of chlorinated phenols present in potable water.

Many studies concerning the ozonation of aromatic compounds have focused on means of opening the aromatic rings.[5-10] However, little work has been done on polymerization by ozone. Gilbert[11] showed that application of low ozone doses to substituted aromatics increased the UV absorption (this phenomenon was attributed to the formation of polyaromatic compounds and quinones). Chrostowski et al.[12] noted that when ozonating phenols, an oxidative coupling mechanism was favored by high-pH and low-ozone doses.

The objective of this study is to compare the polymerization effects obtained by ozonation or by enzymatic reaction (HRP-H_2O_2) on a chlorinated phenol, the 2,4-dichlorophenol.

MATERIALS AND METHODS

The experimental methods used here follow those previously described by Maloney.[4]

In addition, the ozonation conditions included reactor volumes of 5 and 10 liters, and 1 mg/L or 100 mg/L stock solutions of 2,4-DCP prepared in ultrapure water (Millipore) adjusted to pH 7.5 or tap water (TOC: 1.5–1.6 mg/L; pH 7.5–8, alkalinity 200 mg/L as $CaCO_3$). The ozone concentration in the gas was set at 3 and 10 mg/L of air, and the gas flow rate at 0.5 and 1 liter/min.

Analysis Techniques

High-pressure liquid chromatography (HPLC) on reverse phase (RP-HPLC) was used to identify phenols and separate the products of the reaction. The system included a Dupont 870 pump module, 850 absorbance detector, and a Zorba ODS liquid chromatography column (diameter 4.6 mm, length 25 cm).

A binary gradient (0 to 100% methanol in water over 25 minutes) was used to eluate 2,4-DCP and by-products. Organics were detected by a UV absorption detector. In the case of HRP-H_2O_2 experiments, 2,4-DCP used is radiolabeled on each carbon (Amersham International Pic). Two mL of radiolabeled HPLC fractions were then mixed with 8 mL of Picofluor 30 for counting on a Tricarb 300 C (Packard) liquid scintillation system. Carbon mass balances are then established in the collected fractions.

Gel Permeation Chromatography

The apparent molecular weight distribution is determined by permeation chromatography on Sephadex G25. Conditions are as follows: column (diameter 2.6 cm, length 90 cm) running at 100 mL/hr with nonbuffered ultrapure water (Millipore), pH 5.6 as eluant. Exclusion limits are defined using blue dextran, and standardizations are performed using polyethylene glycols. UV absorption (260 nm) and TOC are measured on the collected fractions.

Chlorine Evolution

Chlorine mineralization was followed by the measurements of the chloride ion concentration and the total organic halide (TOX).

By-Products Identification

By-products identification was done using:

- a gas chromatograph (Carbo Erba 4160) coupled with a mass spectrometer (Ribermag R10–10C) using three types of capillary columns (OV 1701, Cp Sil 5, SPS5) with full-spectrum electron impact at 70 eV
- direct introduction into the mass spectrometer using two different techniques: positive or negative chemical ionization desorption (VG ZABHT) or electronic impact desorption (Ribermag R10–10C)

RESULTS AND DISCUSSION

Elimination of 2,4-DCP

Whatever the initial concentrations of 2,4-DCP (1 mg/L and 100 mg/L), the ozonation conducted at pH 7.5–8 removed more than 90% of 2,4-DCP. This removal was obtained for introduced ozone-carbon quantity ratio equal to 1.3 mg O_3/mg C (Figure 1). This ratio corresponded to ozone doses generally used during preozonation.

2,4-DCP was efficiently removed by the enzymatic system coupled with peroxidase-hydrogen peroxide. Tests conducted with initial concentrations of 2,4-DCP ranging from 2.5 μg/L to 240 mg/L showed that after 3 hr of reaction time and using a low enzymatic activity-initial concentration of 2,4-DCP ratio, the removal was better than 95% (Figure 2, Table 1).

These two processes studied were very efficient, but the reduction pathway must be determined: polymerization or ring opening? While these studies demonstrated the efficiency of the processes, they did not address the mechanism. The removal pathway (polymerization vs ring opening) must be determined.

Evolution of Global Parameters

Ozonation of 2,4-DCP was characterized by three main modifications:

- The formation of compounds that absorb at 420 nm. This absorption had a maximum and decreased after the total reduction of 2,4-DCP (Figure 3).

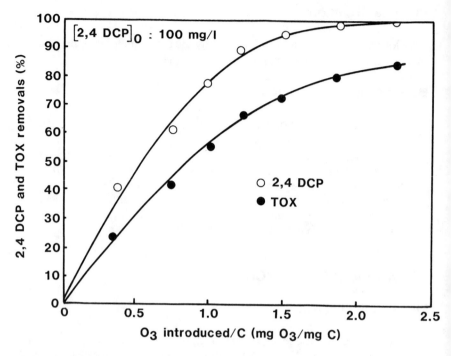

Figure 1. Evolution of 2,4-DCP and organic chlorine removals versus the applied ozone dose in tap water. $[2,4\text{-DCP}]_0 = 100$ mg/L.

- The removal of organic chlorine and subsequent formation of chloride ions. Destruction reached 80% when 2,4-DCP was totally eliminated (Figure 1).
- The formation of "insoluble" compounds retained on 0.4 μm membrane. The insoluble TOC quantity depended on the introduced ozone dose (Figure 3). The percentage increased from 25% to 40% when the introduced ozone-carbon quantity ratio was increased from 1 to 2.3 (Figure 3). Insoluble compounds were also present when tests were run at initial 2,4-DCP concentration equal to 1 mg/L.

Mass balances using radiolabeled carbon have been calculated for enzymatic reaction at different initial concentrations of 2,4-DCP. These calculations showed that the quantity of insoluble compounds was of the same order as those obtained after ozonation (Figure 4). Loss of organic chlorine was, in this case, less important; only 35% of organic chlorine loss was measured as chloride ions when 2,4-DCP was entirely eliminated.

While these results showed the formation of polymers that were partially insoluble, they did not give any information about the degree of polymerization of the soluble fraction. This parameter could be estimated by complementary experiments.

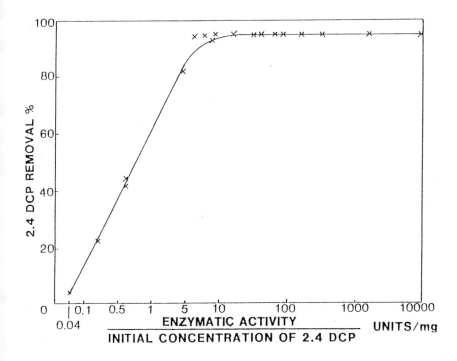

Figure 2. Evolution of 2,4-DCP removal versus the enzymatic activity. $[H_2O_2]_0$ = 170 mg/L; pH 7; reaction time 3 hr.

Table 1. Removal of 2,4-DCP: Influence of 2,4-DCP Initial Concentrations and Enzymatic Activities[a]

2,4-DCP (mg/L)	% Removal of 2,4-DCP		
	Enzymatic Activity (units/L)		
	1000	100	10
240	—	47–42	7–10
120	93–95	—	
60	> 95	94–92	25–22
24	> 95	95	47–50
10	> 95	> 95	—
3	> 95	> 95	82–80
0.6	> 95	> 95	> 95
0.06	> 95	> 95	> 95
0.008	> 90	> 90	—
0.0025	> 90	> 90	—

[a](H_2O_2) = 170 mg/L; ultrapure water and tap water, pH 6.5–7.

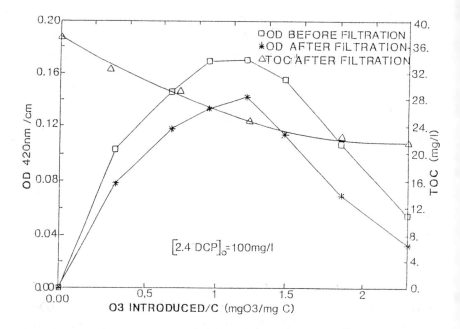

Figure 3. Evolution of absorption (420 nm) and TOC versus the applied ozone dose. [2,4-DCP]$_0$ = 100 mg/L.

Apparent Molecular Weight Distribution of Oxidation By-Products

The solutions of 2,4-DCP (1 mg/L and 100 mg/L) were either ozonated or reacted with horseradish peroxidase. The solutions were then filtered on 0.45 μm membrane, and the exclusion chromatograms were established for the different filtrates after concentration. Exclusion chromatograms obtained for tests using initial concentration of 2,4-DCP equal to 100 mg/L showed the formation of four main peaks, corresponding to apparent molecular weights above 5000 daltons (peak a), between 1000 and 5000 daltons (peak b), and below 1000 daltons (peaks c and d) (Figures 5 and 6). Some carboxylic acids were formed during ozonation even at low ozone doses. If, when ozonated samples were tested by exclusion chromatography, carboxylic acids would be eluted more rapidly or retained on gel, in the first case carboxylic acids would contribute to the fractions of higher-molecular-weight compounds. To verify this possible effect, solutions of different carboxylic acids were tested by exclusion chromatography running in the same conditions. It was found that maleic, fumaric, malonic, oxalic, and acetic acids were well eluted in the lower-molecular-weight fractions (c,d).

Chromatograms showed also that more than 90% of TOC from soluble fraction contained compounds of molecular weights below 5000 daltons. The

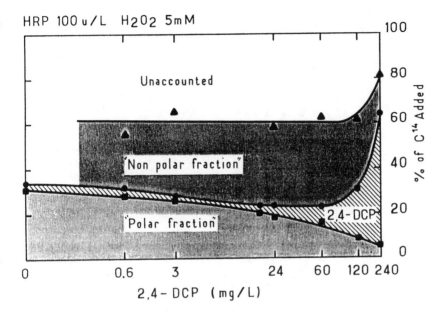

Figure 4. Distribution of radiolabeled HPLC fractions versus initial 2,4-DCP concentrations. HRP: 100 units/L; $[H_2O_2]_0$ = 170 mg/L; pH 7.

increase of the applied ozone dose raised significantly the fraction of insoluble compounds (Table 2).

The results obtained in the case of ozonation conducted with an initial 2,4-DCP concentration equal to 1 mg/L showed a shift in the peak toward lower molecular weights, since the main peak corresponded to peak c (Figure 5).

If exclusion chromatography demonstrated the degree of polymerization of the oxidation by-products, a better explanation might be obtained with complementary analysis.

Distribution of Oxidation By-Products in Terms of Polarity

In each test, the evolution of 2,4-DCP concentration was followed by reverse-phase HPLC. This technique also showed the formation of by-products classified as a function of the polarity. Two main classes might be defined: more or less polar than 2,4-DCP. The comparison of chromatogram profiles (Figures 7 and 8) showed a large difference between the two processes: ozonation produced primarily compounds more polar than 2,4-DCP, while enzymatic reaction led to "nonpolar" polymers. However, the detection of these compounds was based on UV absorption. These profiles corresponded only to compounds that absorbed in the UV range. To correct this,

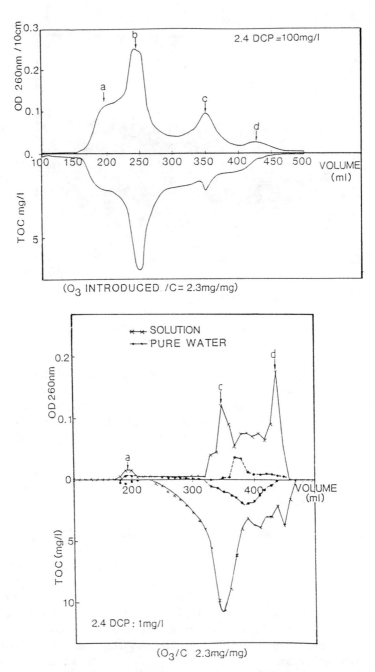

Figure 5. Exclusion chromatogram (Sephadex G25) of ozonated 2,4-DCP solution. [2,4-DCP]$_0$ = 100 mg/L, and 1 mg/L; O$_3$ introduced/C = 2.3 mg/mg.

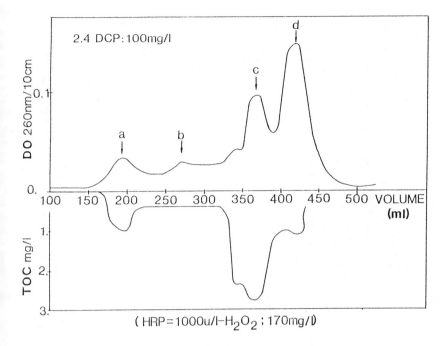

Figure 6. Exclusion chromatogram (Sephadex G25) of 2,4-DCP solution after an enzymatic reaction. $[2,4\text{-DCP}]_0 = 100$ mg/L; HRP: units/L; $[H_2O_2]_0 = 170$ mg/L.

in the case of enzymatic experiments, mass balances on radiolabeled carbon were calculated.

Figure 4 gives an idea of results obtained after enzymatic reaction for different initial concentrations of radiolabeled 2,4-DCP. The nonpolar fraction was the main fraction, and the polar fraction represented about 20% of initial TOC.

This characterization by polarity is important and gives some explanations for the mass spectra interpretation. On the other hand, polarity is an impor-

Table 2. Apparent Molecular Weight Distribution of Oxidation By-Products Obtained by Exclusion Chromatography[a]

	Insoluble Fraction (% TOC)	>5000 (a) (% TOC)	1000–5000 (b) (% TOC)	>1000 (c) (% TOC)	(d) (% TOC)
$O_3/C = 1$	25	14	37	10	14
$O_3/C = 2.3$	40	9.0	30	8	13
$HRP–H_2O_2$	45	10	10	27	8

[a]% TOC with regard to the initial TOC of the 2,4-DCP solution. $[2,4\text{-DCP}]_0 = 100$ mg/L.

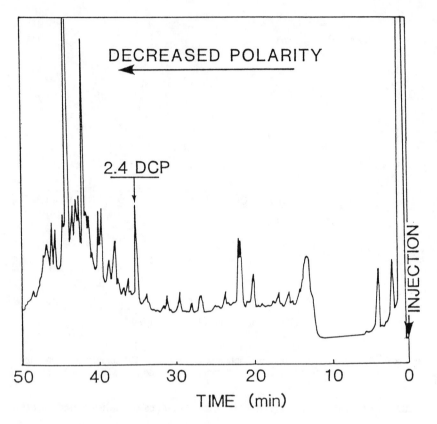

Figure 7. HPLC chromatogram of radiolabeled 2,4-DCP solution after an enzymatic reaction ($[2,4\text{-DCP}]_0 = 100$ mg/L).

tant parameter in the choice of the subsequent treatment process (adsorption and so on).

Identification of By-Products by Mass Spectrometry

Mass spectrometry (MS) (sometimes coupled with gas-phase chromatography) was used to identify by-products in the different fractions.

Direct Introduction

Electron-impact spectra obtained after the direct introduction of insoluble matter exhibited mass fragments between 300 and 1000. This corresponded to dimeric to hexameric molecules (Figure 9). Higher-molecular-weight polymers could not be identified for two main reasons: the sensitivity of the

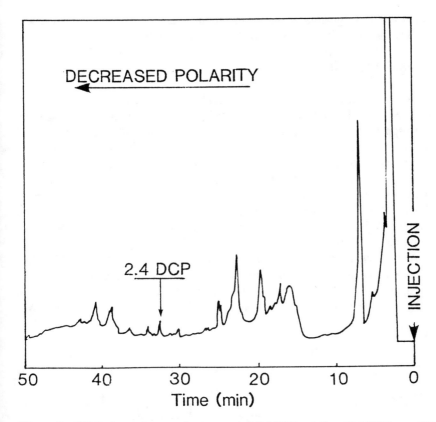

Figure 8. HPLC chromatogram of an ozonated 2,4-DCP solution ($[2,4\text{-DCP}]_0$ = 100 mg/L; O_3 introduced/C = 1 mg/mg).

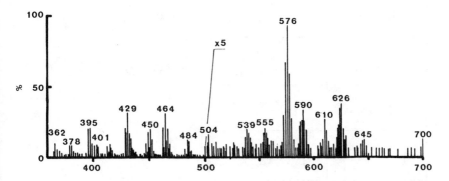

Figure 9. Direct-introduction electron-impact mass spectrum of the insoluble fraction obtained after enzymatic treatment of a 2,4-DCP solution.

apparatus to fragments of higher molecular weight was limited, and it was difficult to volatilize highly polymerized compounds.

GS-MS Analysis

Different fractions (soluble, insoluble) were extracted by dichloromethane or preseparated by HPLC, before analysis by GC-MS. Structures compatible with spectra obtained after enzymatic reaction and ozonation are summarized in Table 3. These interpretations were partly based on the examination of chlorinated isotopic clusters detected by soft ionization techniques (positive or negative chemical ionization) carried out on preseparated fractions. Also, the compatibility of the mass spectra with the possible reaction products of 2,4-DCP was considered in this interpretation. These compounds resulted from oxidative coupling of 2,4-DCP by means of C—O—C and C—C bonds (dioxins and chlorinated dibenzofurans).

After each type of oxidation, many compounds were found which had masses equal to 304 and 322 (Figure 10) but some remained unidentified. As demonstrated before, some identified compounds had aromatic rings that have lost one or two chlorine atoms. In the case of ozonation, the GC-MS

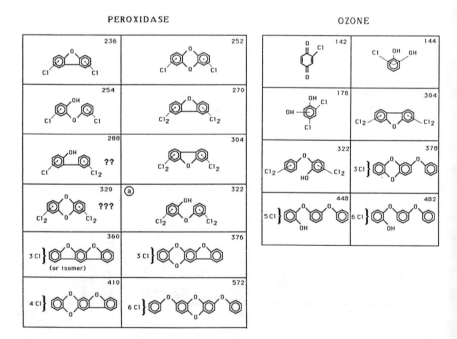

Figure 10. Formulas of oxidation by-products identified by mass spectrometry.

identified aromatic compounds which were entirely dechlorinated (masses 268, 324, 330, 340, and 386) or dihydroxybenzenes with or without the loss of one chlorine atom (masses 144 and 178).

But even if the coupling pathway is known (C—C and C—O—C bonds between rings), exact structures of higher polymers remain to be determined.

CONCLUSIONS

Whatever the initial concentrations of 2,4-DCP, the elimination is better than 95% for low-ozone doses or enzymatic activities corresponding to small quantities of enzyme (10 units/mg of DCP). Under these conditions, 2,4-DCP is removed following an oxidative coupling pathway as shown by different analysis techniques. This polymerization is characterized by the formation of quantities of insoluble compounds (25% to 45% for 100 mg/L initial concentration of 2,4-DCP) easily removed by classical treatments, and soluble compounds which have apparent molecular weights, for the most part, below 5000 daltons.

If polymerization occurs by ozonation or enzymatic reaction, some differences are noted:

- Enzymatic reaction produces a higher quantity of polymers that have molecular weights below 1000 daltons.
- UV absorption by-products obtained by ozonation are generally more polar than those obtained by enzymatic reaction.
- Dechlorination is more important in the case of ozonation.

Mass spectrometry used with and without gas-phase chromatography showed, in all fractions, polymers of at least hexamer size (the technique used limits the identification of higher polymerization degree) and chlorinated polyaromatic compounds that in some cases lost one or two chlorine atoms per ring.

These oxidative techniques produced many identical compounds even at low initial concentrations of 2,4-DCP.

These first results indicate that polymerization techniques may be useful water treatment methods, but many questions must be resolved concerning the influence of the organic matrix on the removal efficiency, the nature of by-products, the evolution of short- and long-term toxicities levels, the detection of newly formed compounds, and the optimization for the removal of the polymerization by-products by subsequent treatment such as coagulation.

The development of advanced ozonation techniques such as the combinations O_3 + UV or O_3 + H_2O_2, mainly based on the hydroxyl radical reactivity, puts the question of this oxidant species effect on polymerization efficiency. Experiments are being conducted and the mechanistic approach will be presented in a paper to be published.

REFERENCES

1. Alberti, B. N, and A. M. Klibanov. "Enzymatic Removal of Dissolved Aromatics from Industrial Aqueous Effluents," *Biotechnology and Bioengineering Symposium, no. 11,* (New York: John Wiley & Sons, 1981), 373–379.
2. Klibanov, A. M., and E. D. Morris. "Horseradish Peroxidase for the Removal of Carcinogenic Aromatic Amines from Water," *Enzyme Microb. Technol.* 3:119–122 (1981).
3. Klibanov, A. M., B. N. Alberti, E. D. Morris, and L. M. Felshin. "Enzymatic Removal of Toxic Phenols and Anilines from Wastewaters," *J. of Appl. Biochem.* 2:414–421 (1980).
4. Fiessinger, F., S. W. Maloney, J. Manem, and J. Mallevialle. "Potential Use of Enzymes as Catalysts in Drinking Water for the Oxidation of Taste Causing Substances," *Aqua* 2:116–118 (1984).
5. Doré, M., B. Langlais, and B. Legube. "Mechanism of the Reaction of Ozone with Soluble Aromatic Pollutants," *Ozone Sci. Eng.* 2:39–54 (1980).
6. Gilbert, E. "Über den Abbau von Organischen Schädstoffen in Wasser durch Ozon," *Vom Wasser* 43:275–290 (1974).
7. Mallevialle, J. "Action de l'ozone dans la dégradation des composés phénoliques simples et polymérisés: Application aux matirès humiques contenues dans les eaux," *T.S.M. L'Eau* 3:107–113 (1975).
8. Yamamoto, Y., E. Niki, E., H. Shiokawa, H., and I. Kamiya. "Ozonation of Organic Compounds 2. Ozonation of Phenol in Water," *J. Org. Chem.* 44:2137–2142 (1979).
9. Li, K., C. H. Kuo, and J. L. Weeks. "A Kinetic Study of Ozone-Phenol Reactions in Aqueous Solution," *Am. Inst. Chem. Eng. J.* 25:583–591 (1979).
10. Singer, P. C., and M. D. Gurol. "Dynamic of the Ozonation of Phenol. I: Experimental Observations," *Water Res.* 17(9):1163–1171 (1983).
11. Gilbert, E. "Investigations on the Changes of Biological Degradability of Single Substances Induced by Ozonation," *Ozone Sci. Eng.* 5:137–149 (1983).
12. Chrostowski, P. C., A. M. Dietrich, and I. H. Suffet. "Ozone and Oxygen Induced Oxidative Coupling of Aqueous Phenolics," *Water Res.* 17(11):1627–1633 (1983).

CHAPTER 15

By-Products from Ozonation and Photolytic Ozonation of Organic Pollutants in Water: Preliminary Observations

Gary R. Peyton, Chai S. Gee, and Michelle A. Smith, Illinois State Water Survey, Champaign, Illinois

John Bandy and Stephen W. Maloney, U.S. Army Construction Engineering Research Laboratory, Champaign, Illinois

INTRODUCTION

The so-called advanced oxidation processes such as ozonation, photolytic ozonation, O_3/H_2O_2, and H_2O_2/UV are of great interest in water treatment because of their ability to completely "mineralize" organic pollutants to carbon dioxide and water.[1-5] However, little is known concerning the intermediate by-products of these reactions. Treatment situations sometimes arise in which it would be desirable to destroy some highly toxic parent compound but impractical to eliminate all organic material from solution. It is therefore of interest to identify by-products from a given parent compound in order to assess the potential toxicity of the mixture which results from partial treatment.

Identification of oxidation by-products is also extremely important from the standpoint of developing a predictive model of the treatment processes. Such a model for photolytic ozonation is under development in the Oxidation Research Laboratory at the Illinois State Water Survey. This model will be described below, and the feedback relationship between the model development and product identification studies will be described.

BACKGROUND

Several studies of ozonation by-products have been performed in the past. A review of the literature up to 1980 is given in reference 6. More recently, the works of Gilbert et al.[7] and references therein and of Huang[8] are notable for the extent to which some rather uncommon by-products have been identified. However, few studies have come close to so completely identifying the entire suite of oxidation products that the entire degradation pathway could be written with any confidence. One reason for this problem is that GC/MS, the usual method of choice for product identification, is not particularly suited for identification of many of the labile oxidation intermediates, because of their thermal instability. Gentler methods such as HPLC are needed to avoid altering the by-products during the analytical process.

The most important reactions that occur in the oxidation systems named above are those of ozone and hydroxyl radical. Ozone reacts with double bonds to form a primary ozonide, which decomposes to a carbonyl compound and a carbonyl oxide zwitterion (I). The latter is intercepted in aqueous solution before the two products can recombine to form a secondary ozonide.[9]

$$^-O - O - {}^+CR_1R_2 + H_2O \longrightarrow HO_2CR_1R_2 \quad (1)$$
$$\text{(I)} \qquad\qquad\qquad\qquad \overset{\displaystyle OH}{\underset{\displaystyle \text{(II)}}{|}}$$

The resulting hydroxyhydroperoxide (II) may eliminate hydrogen peroxide to become a second carbonyl compound.

Hydroxyl radicals react with saturated aliphatic compounds by hydrogen abstraction. The resulting organic radical reacts quickly with oxygen, which is always present in these systems, to form a peroxy radical, $RO_2 \cdot$, which may then abstract a hydrogen atom to become RO_2H or react with another peroxy radical to give a tetroxide.[10] Although the aqueous chemistry which follows is less well understood, the decomposition products of these intermediates are organic acids and carbonyl compounds.[10-12] Hydroxyl radicals react with aromatic molecules to eventually cleave the ring and yield unsaturated acids and carbonyl compounds.[13]

The common thread which runs throughout these degradations is the observation of organic acids and particularly carbonyl compounds as stable products. Derivatization of carbonyl compounds using 2,4-dinitrophenylhydrazine (DNPH) followed by HPLC separation and UV/visible absorption detection seemed to be an ideal method of carbonyl compound analysis because of the low water solubility, high extinction coefficient, and supposed stability of the derivatives. There are many papers in the literature which describe the use of some variant of this method. Some inconsistency was found in the environmental literature, however, concerning results obtained using this method. Some investigators reported difficulty in analysis for common simple compounds with which other investigators apparently had no difficulty at all. This inconsistency prompted us to further investigate the stability of DNPH derivatives.

It is well-known[14] that carbonyl compounds react with hydrogen peroxide to form hydroxyhydroperoxides. The latter reach equilibrium with parent compounds as well as more complex peroxidic substances. Hydrogen peroxide is almost always present in the advanced oxidation process systems, as are carbonyl compounds. It is therefore of interest to determine the quantities and lifetimes of peroxidic species which may be formed. The speed with which equilibrium is attained has a bearing not only on health effects due to products present in the treatment/distribution system, but also on the analytical chemistry which must be used to detect such intermediates.

The preliminary studies described here were carried out as part of a study to determine 1) if peroxidic compounds can be expected to be present in appreciable quantities as a result of the advanced oxidation processes, and 2) if identifiable oxidation by-products found in photolytic ozonation experiments are consistent with those predicted by the mechanistic model at its present stage of development and current knowledge of aqueous hydroxyl radical chemistry.

EXPERIMENTAL

Reactor, Ozone Manifold, and Reactor Runs

The stirred-tank photochemical reactor and the ozone generating and monitoring system have been described earlier.[15] Aqueous solutions of the model compounds were treated by photolytic ozonation in the stirred-tank photochemical reactor. Samples were withdrawn as a function of time during treatment, and carbonyl compounds derivatized using 2,4-dinitrophenylhydrazine (DNPH) as described below. Samples were also withdrawn and analyzed for organic acids, ozone, hydrogen peroxide, and the pH measured.

Analytical Procedures

Carbonyl Compounds

Ten milliliters of the aqueous solution of carbonyl compound were pipetted into a 20-mL extraction vial and sparged with nitrogen for 10 minutes. Two mL of DNPH reagent (0.25% 2,4-dinitrophenylhydrazine in 6 N HCl) was added, the mixture shaken vigorously, sparged again briefly, and allowed to stand at room temperature overnight. Earlier studies had shown that a reaction time of only a few hours was not adequate for some carbonyl compounds. After standing, the solution was extracted with 2 mL of nitrogen-sparged methylene chloride. This phase was separated from the aqueous phase and 1.5 mL of sparged dimethylformamide (DMF) added to the organic phase. Since methylene chloride was insoluble in the HPLC mobile phase (initially 50% DMF in water), the methylene chloride was evaporated at 60°C under a stream of nitrogen. If necessary, the remaining solution was made back up to exactly 1.5 mL with DMF, and analyzed by HPLC on a 25-cm Supelco LC-DP (diphenylmethyl silyl) column, using UV absorption detection at 360 nm. Retention volumes and peak areas were compared with DNPH derivatives prepared from authentic samples of carbonyl compounds.

Organic acids were determined as the benzyl esters by adjusting the aqueous solution to pH 8 with tetrabutylammonium hydroxide, blowing down to dryness with nitrogen at 60°C, and taking the residue up in dry acetone containing a slight excess of benzyl bromide, using propionic acid as a surrogate recovery standard. The mixtures were heated at 60°C for one hour then cooled and analyzed by GC/FID or capillary column GC/MS.

Aqueous ozone and hydrogen peroxide analyses were by the indigo and titanium methods described in earlier publications.[15,16]

RESULTS

Formation of Hydroxyhydroperoxides from Hydrogen Peroxide and Carbonyl Compounds

The presence of peroxides, hydroperoxides, and hydroxyhydroperoxides in treated water is of potential concern from the point of view of health effects. The reaction between formaldehyde and hydrogen peroxide was followed in order to determine how fast the equilibrium between hydrogen peroxide and a typical oxidation product was established, and to what extent the reaction proceeded.

Hydrogen peroxide (1 mM) was mixed with 1 mM formaldehyde and the apparent hydrogen peroxide concentration monitored as a function of time. This data is shown in Figure 1, where it is seen that there is a sharp decrease in H_2O_2 concentration immediately after mixing with formaldehyde. Upon standing with the H_2O_2 reagent, color develops further, indicating that H_2O_2 is being liberated in solution, since the colorimetric reaction is specific for hydrogen peroxide. As the H_2O_2/formaldehyde mixture ages, a steady decrease in both instantaneous and total available H_2O_2 is seen with time. Finally, the regrowth of H_2O_2 concentration measured in the 100-minute sample was measured to determine the time scale of peroxide regeneration.

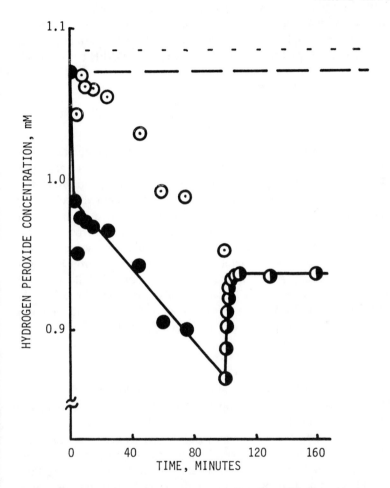

● INITIAL ABSORBANCE

◉ FINAL ABSORBANCE

◑ DEVELOPMENT OF FINAL
     ABSORBANCE FOR 100
     MINUTE SAMPLE

Figure 1. Formation of hydroxyhydroperoxide from formaldehyde and hydrogen peroxide.

Oxidation of DNPH Derivatives

The effect of the presence of oxygen on the DNPH derivatization of carbonyl compounds was studied to determine whether an inert atmosphere was required during derivatization. The DNPH derivatization method described in the experimental section was carried out on aliquots of standard solutions of carbonyl compounds. Other samples were taken through the procedure in a parallel fashion except that the solutions were sparged with oxygen between each step rather than with nitrogen. The results for acetone are shown in Figure 2, where it is seen that the peak (peak A) corresponding to the analyte in the chromatogram from the nitrogen-sparged sample is completely absent in the oxygen-sparged sample. Instead, several other peaks are seen which were not present when the sample was kept free of oxygen. From these results it was concluded that it was important to eliminate oxygen from the reaction mixture while preparing DNPH derivatives.

Photolytic Ozonation Experiments

Acetate

Solutions of acetic acid (3 and 4.7 mM) were treated by photolytic ozonation using 3½ lamps—four low-pressure mercury arc lamps (one half-covered)—rated at 5½ W output apiece at 254 nm and an applied ozone dose of 3 liters/minute of 1.5% ozone. The quantitative data from two preliminary experiments are not in agreement and will not be reported here. Some general observations will be made, however, which are common to both experiments. Glyoxylic acid and formaldehyde were found in one experiment where catalase was added immediately to the samples to destroy hydrogen peroxide. In the other experiment where catalase was not added, no glyoxylic acid was found. There was only a trace of glycolic acid, far below the quantifiable limit (<0.05 mM) in both experiments. Glycolic acid would be expected to be stable in the presence of hydrogen peroxide.

t-Butanol

Only the oxidants and carbonyl compounds were followed in this experiment. The evolution of carbonyl compounds is shown in Figure 3, where it is seen that the two initial products are acetone and formaldehyde. Glyoxal appears to form almost from the beginning of the experiment, but either is a secondary product or represents a parallel pathway. Glyoxylic acid and an unidentified compound are clearly secondary products, as shown by their appearance at the time that production of the primary products begins to diminish. The fact that formaldehyde does not disappear as quickly as acetone in the later stages of the reaction probably indicates that it is also being produced by another pathway at that point.

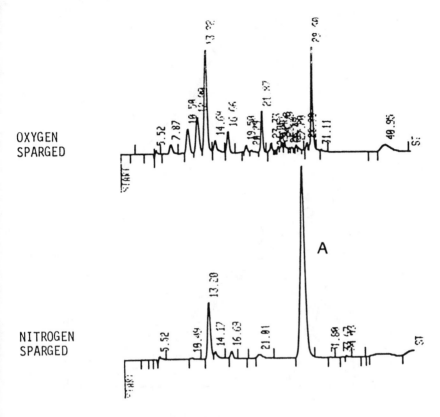

Figure 2. Effect of oxygen on 2,4-dinitrophenylhydrazine derivative of acetone. Shown are HPLC chromatograms with UV absorbance detection at 360 nm. Peak A is the analyte of interest.

DISCUSSION

Oxidation of DNPH Derivatives

The results of the N_2/O_2 sparge experiments show that it is important to protect samples from oxygen during and after derivatization with DNPH. No mention of this problem has been found in either the environmental or organic analysis literature by these investigators, although papers in which DNPH has been used to analyze carbonyl compounds in environmental samples are published regularly in the environmental literature. The previously discussed results concerning the carbonyl/H_2O_2 reaction suggest that it may also be desirable to destroy hydrogen peroxide in samples before attempting carbonyl derivatization. This question is currently under investigation.

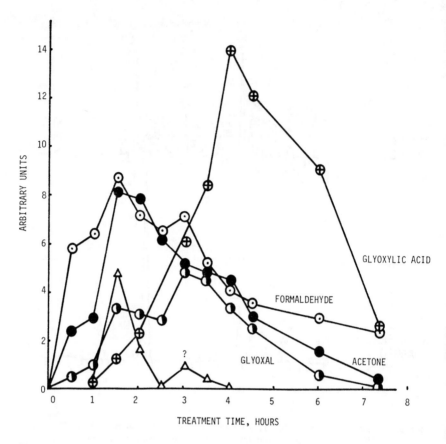

Figure 3. Products from photolytic ozonation of *t*-butanol.

Formation of Hydroxyhydroperoxides

The results in Figure 1 are interpreted in terms of an equilibrium,[14] which is observable on the laboratory time scale, between formaldehyde and hydrogen peroxide, on the one hand, and hydroperoxide (equation 2).

$$CH_2O + H_2O_2 \rightarrow \underset{(I)}{HOCH_2O_2H} \rightarrow \overset{\text{other}}{\text{hydroperoxides}} \qquad (2)$$

When reagent is added, only H_2O_2 reacts and the lower "instantaneous" concentration is detected. Upon standing, H_2O_2 is regenerated from species I. The amount of "bound" H_2O_2 is seen to be a significant fraction (about 10%). With time, however, one of the species participating in the equilibrium is destroyed, leading to the net decrease in total H_2O_2 seen in Figure 1. Monitoring the H_2O_2 regeneration rate for the 100-minute sample shows that the equilibration is complete in 5–10 minutes.

In municipal water treatment systems, the implication is that any reaction which removes H_2O_2 or carbonyl compound, either in postozonation treatment or during distribution, will eliminate hydroxyhydroperoxide. It is not presently known if the products of the reaction that depletes the equilibria system are themselves peroxidic. However, there is a stronger implication for point-of-use oxidation systems, which are currently under consideration and commercially available for residential use. Even assuming that the nonequilibrium products are not peroxidic, the time-scale of the degradation of peroxide is such that a storage time would be required for its depletion, which may be long compared to that being considered for point-of-use treatment/immediate-consumption systems. The question of formation and degradation of these peroxidic compounds under actual drinking water treatment conditions requires closer study.

Degradation Pathways of Organic Compounds Upon Photolytic Ozonation

The core of the mechanistic model for the ozone/UV/hydrogen peroxide system has been presented earlier[15–18] and is given in Figure 4. Direct reaction of organic compound with ozone has been omitted from this system for clarity, but is important for many cases and must be included in the mass balance considerations. As described previously,[15] this model has been assembled from the work of Staehelin and Hoigne, Rabani et al., and others, along with our earlier work on ozone photolysis. A complete listing of the important references can be found in reference 15 or 17 of this chapter.

For organic substrates HRH (Figure 4) which produce peroxyl radicals ($HRO_2\bullet$) capable of eliminating superoxide (O_2^-), a cyclic reaction along the inner circle of Figure 4 is quickly established. The quantitative utility of this oxidant-balance model for compounds whose peroxy radicals can easily eliminate superoxide has been demonstrated using the methanol/formaldehyde/formic acid system.[19] When the radical $HRO_2\bullet$ cannot release superoxide, however, the reaction pathways are more complicated ($HRO_2 \rightarrow H_2O_2$, upper left portion of Figure 4). The two compounds chosen for this study are examples of compounds whose peroxy radicals do not easily eliminate superoxide. Acetic acid is, in fact, the smallest oxygen-containing compound that produces a peroxyl radical that will not eliminate superoxide directly.

It is significant that no substantial amounts of four-carbon carbonyl compound (i.e., no unidentifiable carbonyl compound) was found as a primary product in the photolytic ozonation of t-butanol. This implies that a carbon-carbon bond was broken before the first major stable products were formed.

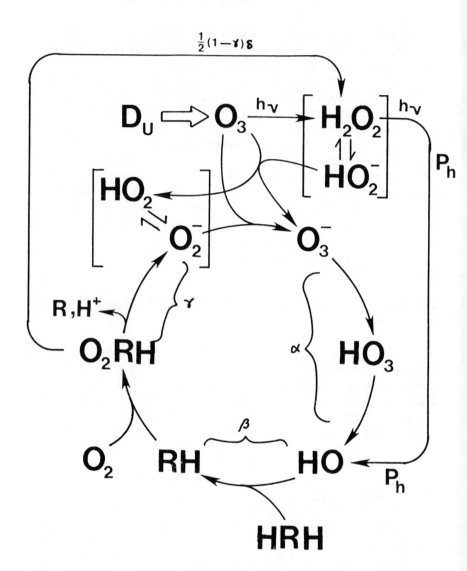

Figure 4. Mechanistic model for photolytic ozonation. D_u is the "utilized" ozone dose, i.e., that which is transferred out of the gas phase.

Both of these observations are consistent with the formation of an intermediate tetroxide species.[10,11] This species is thought to be formed by the following process: hydrogen abstraction from the parent compound (equation 3) by hydroxyl radical yields an organic radical that quickly reacts with dioxygen to form a peroxy radical (equation 4).

$$HRH + \cdot OH \rightarrow HR\cdot + H_2O \tag{3}$$

$$HR\cdot + O_2 \rightarrow HRO_2\cdot \tag{4}$$

Unable to eliminate superoxide directly, $HRO_2\cdot$ reacts with another similar radical, forming the tetroxide (equation 5).

$$2HRO_2\cdot \rightarrow HRO_4RH \tag{5}$$

The tetroxide can then decompose to form products R, HRO, O_2, H_2O_2, etc.,

$$HRO_4RH \rightarrow 2RO + H_2O_2 \tag{6}$$

$$HRO_4RH \rightarrow 2HRO + O_2 \tag{7}$$

depending on the structure of R.

Some possibilities for the decomposition of the tetroxide formed from peroxyacetate are shown in Figure 5. In addition to the intermediate configurations shown in that figure, there is another intermediate geometry in which the two "sides" of the tetroxide interact by hydrogen bonding of the alpha hydrogen on one side to the carbon-bonded oxygen of the other side. This pathway, known as the Russell mechanism,[11] leads to the production of both glyoxylic and glycolic acids, and the generation of oxygen, rather than hydrogen peroxide, as shown in Figure 6. As the latter of those acids was not found in appreciable quantities, it is assumed that the Russell mechanism does not contribute greatly to the mechanism in the present system.

The results do not seem to correspond completely to those found by von Sonntag et al.[20] for hydroxyl radical reactions produced by the pulse radiolysis of N_2O/O_2-saturated aqueous acetate solutions. Those investigators' product analysis led them to conclude that a mixture of mechanisms was operative. There may be differences in the chemistry of the two systems (pulse radiolysis versus ozone/UV), perhaps due to the presence of ozone in our study.

Some possible intermediates in the decomposition of the tetroxide formed from the peroxy radical from t-butanol are shown in Figure 7. Both of these possibilities, i.e., hydrogen bonding of the methyl or hydroxyl hydrogens to the tetroxide bridge, lead to the same products: acetone, formaldehyde, and hydrogen peroxide. The Russell mechanism and other possibilities would lead to the formation of alpha-hydroxyisobutyraldehyde, which was not detected as a primary product. Schuchmann and von Sonntag[21] found both alpha-hydroxyisobutyraldehyde and 1,2-dihydroxy-2-methylpropane as a result of pulse radiolysis of aqueous t-butanol solutions. Our analytical methods would not have detected the latter compound. Ozone utilization data

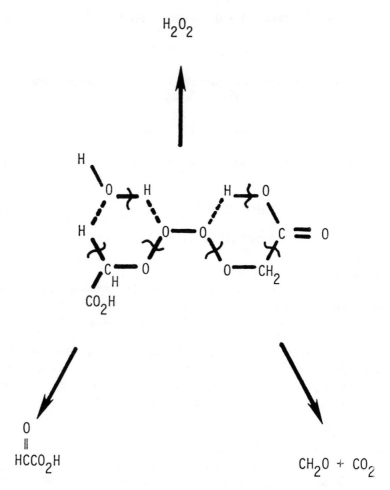

Figure 5. Possibilities for decomposition of tetroxide formed from peroxyacetate.

cannot yet be used in the mechanistic argument due to the necessity of identifying all stable intermediates. We thus tentatively conclude, in agreement with Schuchmann and von Sonntag,[20,21] that the tetroxide mechanism is consistent with the experimental data, since it neatly explains the scission of the carbon-carbon bond and formation of formaldehyde from both t-butanol and acetate. Formation of methyl radical by reaction 8,

$$CH_3CO_2^- + \cdot OH \rightarrow CH_3\cdot + CO_2 + {}^-OH \tag{8}$$

followed by reactions 4, 5, and then the Russell reaction (Figure 6), or decomposition of the methylperoxy tetroxide by a pathway analogous to the

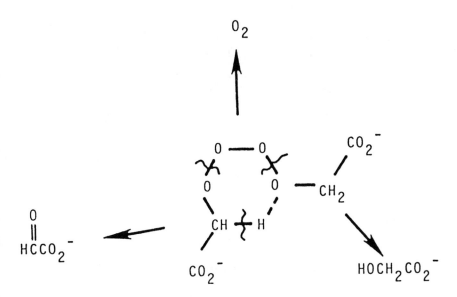

Figure 6. Russell mechanism applied to peroxyacetate tetroxide.

left-hand side of Figure 5, however, also yields formaldehyde, and cannot be ruled out on the basis of our data. It has been shown, however, that in pulse radiolysis studies, the reaction of hydroxyl radical with acetic acid proceeds almost entirely by alpha-hydrogen abstraction.[22]

CONCLUSIONS

1) Carbonyl compounds formed as products of oxidation reactions can react with hydrogen peroxide, which is frequently present in such systems, to establish an equilibrium between parent compounds, significant amounts of hydroxyhydroperoxide, and more complex peroxidic products. The results indicate that peroxidic products should be looked for in reaction mixtures from advanced oxidation processes. Since the time scale of this equilibrium is on the order of a few minutes, special precautions are necessary to detect these intermediate products. There may be health implications, particularly in cases where drinking water would be consumed immediately following treatment, as might be the case in a residential point-of-use treatment system.

2) Phenylhydrazine derivatives of carbonyl compounds are oxygen-sensitive. Appropriate precautions should be taken in the analysis of such compounds.

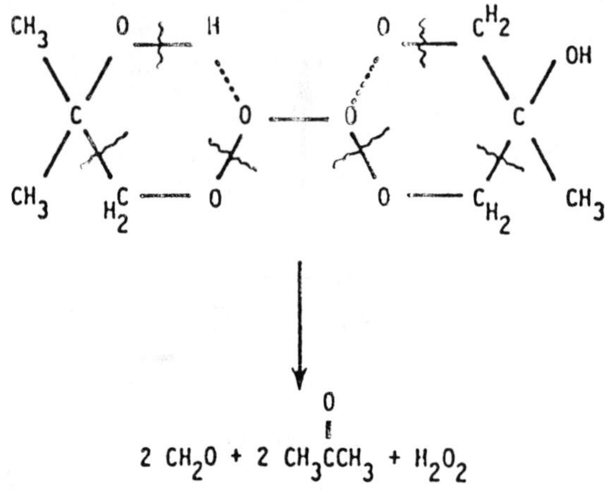

$$2\ CH_2O\ +\ 2\ CH_3\overset{O}{\overset{\|}{C}}CH_3\ +\ H_2O_2$$

Figure 7. Decomposition of *tert*-butanol tetroxide.

3) Preliminary data on two model compounds whose peroxy radicals do not easily eliminate superoxide indicate that the postulated tetroxide intermediate is useful in explaining the stable by-products which are formed in photolytic ozonation experiments.

4) Results of this type provide useful input into the mass balance model for photolytic ozonation.

REFERENCES

1. Glaze, W. H., G. R. Peyton, F. Y. Huang, J. L. Burleson, and P. C. Jones. "Oxidation of Water Supply Refractory Species by Ozone with Ultraviolet Radiation," Final Report, EPA-600/2–80–110, August (1980).
2. Prengle, H., Jr., C. G. Hewes, III, and C. E. Mauk. "Oxidation of Refractory Materials by Ozone with Ultraviolet Radiation," in *Proceedings of 2nd International Symposium on Ozone Technology,* R. G. Rice, P. Pichet, and M. A. Vincent, Eds. (Syracuse, New York: International Ozone Institute, 1975), 224.
3. Fochtman, E., and J. E. Huff. "Ozone-Ultraviolet Light Treatment of TNT Wastewaters," in *Proceedings of 2nd International Symposium on Ozone Technology,* R. G. Rice, P. Pichet, and M. Vincent, Eds. (Syracuse, New York: International Ozone Institute, 1975), 211.
4. Leitis, E., H. Bryan, J. D. Zeff, D. C. Crosby, and M. Smith. "An Investigation into the Chemistry of the UV/Ozone Purification Process," presented at the 4th World Ozone Congress, Houston, TX, November 27–29, 1979.
5. Kuo, P., P. E. S. K. Chian, and B. J. Chang. "Identification of End Products Resulting from Ozonation and Chlorination of Organic Compounds Commonly Found in Water," *Environ. Sci. Technol.* 11:1177 (1977).

6. Rice, R. G., and M. E. Browning. "Ozone for Industrial Water and Wastewater Treatment," EPA-600/2–80–060 (April, 1980). Available from NTIS.

7. Gauducheau, C., E. Gilbert, and S. H. Eberle. "Are the Results of Ozonation of Model Compounds at High Concentrations Transferable to the Conditions of Drinking Water Treatment with Ozone?" *Ozone Sci. Eng.* 8:199–216 (1986).

8. Glaze, W. H., G. R. Peyton, F. Y. Saleh, and F. Y. Huang, "Analysis of Disinfection By-Products in Water and Wastewater," *Intern. J. Environ. Anal. Chem.* 7:143–160 (1979).

9. Howard, J. A. "Free-Radical Reaction Mechanisms Involving Peroxides in Solution," in *The Chemistry of Functional Groups,* Saul Patai, Ed. (New York: John Wiley & Sons, 1983), 700–703.

10. Schulte-Frohlinde, D., and C. von Sonntag. "Radiolysis of DNA and Model Systems in the Presence of Oxygen," in *Oxidative Stress,* Helmut Sies, Ed. (New York: Academic Press, Inc., 1985), 11.

11. Howard, J. A. "Free-Radical Reaction Mechanisms Involving Peroxides in Solution," in *The Chemistry of Functional Groups,* Saul Patai, Ed. (New York: John Wiley & Sons, 1983), 247.

12. Fish, A. "Rearrangement and Cyclization Reactions of Organic Peroxy Radicals," in *Organic Peroxides, Vol. 1,* D. Swern, Ed. (New York: Wiley-Interscience, 1970), 141–194.

13. Balakrishnan, I., and M. P. Reddy. "Mechanism of Reaction of Hydroxyl Radicals with Benzene in the Gamma-Radiolysis of the Aerated Aqueous Benzene System," *J. Phys. Chem.* 74:850–55 (1970).

14. Hiatt, R. "Hydroperoxides," in *Organic Peroxides, Vol. 2,* D. Swern, Ed. (New York: Wiley-Interscience, 1970), 36-38.

15. Peyton, G. R., C.-S. Gee, J. Bandy, and S. W. Maloney. "Catalytic/Competition Effects of Humic Substances on Photolytic Ozonation of Organic Compounds," in *Influences of Aquatic Humic Substances on Fate and Treatment of Pollutants,* I. H. Suffet and P. MacCarthy, Eds. (ACS Advances in Chemistry Series, in preparation).

16. Peyton, G. R., and W. H. Glaze. "Mechanism of Photolytic Ozonation," in *Photochemistry of Environmental Aquatic Systems,* ACS Symposium Series No. 327, R. G. Zika and W. J. Cooper, Eds. (Washington, DC: American Chemical Society, 1987), 76–88.

17. Peyton, G. R., and W. H. Glaze. "Destruction of Pollutants in Water with Ozone in Combination with Ultraviolet Radiation. 3. Photolysis of Aqueous Ozone," in press, *Environ. Sci. Technol.*

18. Glaze, W. H., G. R. Peyton, B. Sohm, and D. A. Meldrum. "Pilot Scale Evaluation of Photolytic Ozonation for Trihalomethane Precursor Removal," Final Report to USEPA/DWRD/MERL, Cincinnati, OH, Cooperative Agreement #CR-808825, J. Keith Carswell, Project Officer, (1984).

19. Peyton, G. R., M. A. Smith, and B. M. Peyton. "A Mechanistic Model of Photolytic Ozonation; Quantitative Verification for Superoxide-Generating Substances," manuscript in preparation.

20. Schuchmann, M., H. Zegota, and C. von Sonntag. "Acetate Peroxyl Radicals, $\cdot O_2CH_2CO_2^-$: A Study on the Alpha-Radiolysis and Pulse Radiolysis of Acetate in Oxygenated Aqueous Solutions," *Z. Naturforsch.* 40b:215–221 (1985).

21. Schuchmann, M., and C. von Sonntag. "Hydroxyl Radical-Induced Oxidation of 2-Methyl-2-Propanol in Oxygenated Aqueous Solution. A Product and Pulse Radiolysis Study," *J. Phys. Chem.* 83:780–784 (1979).
22. Neta, P., M. Simic, and E. Hayon, "Pulse Radiolysis of Aliphatic Acids in Aqueous Solution. I. Simple Monocarboxylic Acids," *J. Phys. Chem.* 73:4207–13 (1969).

CHAPTER 16

Application of Closed Loop Stripping and XAD Resin Adsorption for the Determination of Ozone By-Products from Natural Water

William H. Glaze, Minoru Koga, Edward C. Ruth, and Devon Cancilla, University of California-Los Angeles, Los Angeles, California

INTRODUCTION

The use of ozone in the treatment of drinking water has been increasing as public utilities search for alternatives to the use of chlorine. In the City of Los Angeles a new water filtration plant utilizing ozone has been in operation since 1986. Ozone is used by the plant to improve flocculation and for the control of trihalomethanes (THM).

Ozone is a powerful oxidant and reacts with many of the natural constituents present in water. For example, humic materials produce alkanes, aliphatic aldehydes, ketones, and fatty acids as by-products.[1-4] Much remains to be learned about the chemistry and potential health risks of ozone-produced by-products in drinking water.[5]

In this chapter we describe the results of a study designed to evaluate ozone-produced by-products found in water samples taken from the Los Angeles Aqueduct Filtration Plant. Furthermore, we demonstrate the use of Closed Loop Stripping Analysis (CLSA) and XAD-4/8 resin adsorption methods for the isolation of by-products present in water.

EXPERIMENTAL PROCEDURES

Water Sample

The water samples used in this study were obtained from the Los Angeles Aqueduct Filtration Plant located in the San Fernando Valley, Los Angeles

County, California. Three types of water samples were collected: raw aqueduct water, post-ozone-treated water, and finished water. Finished water had passed through ozone, chemical treatment, filtration, and chlorination. Typical water quality values for this period (February-August 1987) were: total hardness 88–97 mg/L, pH 8.0–8.4, color 2 pcu, total alkalinity 113–123 mg/L as $CaCO_3$, turbidity 0.9–3.4, and TOC 1.1–1.9 mg/L. Chemical doses at the L.A. plant in order of application were: ozone dosage approximately 1 mg/L, ferric chloride 1.9 mg/L, CAT Floc C 2.0 mg/L, and chlorine 1.0–1.5 mg/L. THM levels in the finished water were in the range of 8–22 µg/L.

Laboratory Ozonation

Laboratory ozonations were carried out in a baffled 70-liter (liquid volume) stainless steel continuous stirred tank reactor (CSTR). The CSTR was equipped with a sparging disk at the base for introduction of ozone-containing gases. Collection of off gas was accomplished with ports at the top of the reactor, while similar ports at the reactor's base allowed liquid samples to be removed.

Ozone was generated from high-purity oxygen gas with an OREC generator at a flow rate of 1.0 L/min and a tube current setting that gave ozone dose rates in the range of 4.5–15 mg/min.

CLSA

Water samples were analyzed by the CLSA method described by Grob et al.[6] and modified by Krasner et al.[7] Sample volumes of 800 mL were used for the analysis. Each sample was spiked with three n-alkyl chlorides (C_8, C_{10}, and C_{12}) and 2-ethylhexanal to give a concentration of 100 ng/L for each compound. Samples were stripped at 25 ± 0.5°C for 90 minutes with a flow rate of 1.5 L/min. Stripped organics were trapped on approximately 1.5 mg of granular activated carbon (GAC) in cartridges supplied by the laboratory staff of the Metropolitan Water District of Southern California. The carbon was eluted with approximately 25 µL of carbon disulfide and stored in glass minivials sealed with Teflon®-coated liners.*

The extracts were analyzed as promptly as possible by Gas Chromatography-Mass Spectrometry (GC/MS) using either a Hewlett-Packard Model 5890/5970 GC/MSD or a Finnigan Model 4000/2300 GC/MS with an Incos Data System. Both systems utilized 30 m × 0.25 mm chemically bonded DB-5 fused-silica columns (J & W Scientific). The run conditions were: initial temperature 33°C (1 minute), programming to 92°C at 4 degrees/minute, then programming to 230°C at 10 degrees/minute.

---

*Registered trademark of E. I. du Pont de Nemours and Company, Inc., Wilmington, Delaware.

XAD-4/8 Resin Adsorption Method

Two 70-liter samples of raw aqueduct water were collected in clean glass bottles. The first sample was ozonated at a dose rate of 15 mg/minute for 30 minutes using the laboratory CSTR. Excess ozone was measured using the indigo method[8] and residual ozone was quenched by the addition of a sodium arsenite solution (1.0 w/v %). An internal standard solution containing naphthalene-d_8, anthracene-d_{10}, phenol-d_5, benzoic acid-d_5, phenylethylamine-d_4, and acridine-d_9 each at 570 ng/L, was added to both ozonated and unozonated samples. The samples were then acidified to pH 2 with 18 N H_2SO_4 and passed through XAD-4/8 resin columns (23 mm i.d. × 60 cm) at a flow rate of 450 mL/minute using a stainless steel pump.

The XAD-4/8 resin columns consisted of one hundred milliliters of a 50:50 mixture of XAD-4 and XAD-8 resin. The resins were co-purified by consecutive 24-hr Soxhlet extractions with methanol, acetone-hexane (6:4), and methanol.

The organic compounds adsorbed on the resin were eluted with 200 mL of acetone-hexane (6:4). The acid pH extracts were dried with sodium sulfate, concentrated to 1.0 mL with a K-D evaporator, and methylated using a diazomethane-ether solution. An external standard (2-fluorobiphenyl) was added to the final concentrates before GC/MS analysis to determine the recovery of the internal standards.

XAD-recovered compounds were chromatographed starting at an initial temperature of 35°C (1 minute) going to 200°C at 5 degrees/minute. The oven temperature was then raised to 280°C at 10 degrees/minute and held there for 5 minutes.

RESULTS AND DISCUSSION

CLSA

Figures 1–3 show total ion chromatograms for CLSA extracts of raw aqueduct, postozonation, and finished water from the Los Angeles Aqueduct Filtration Plant. Table 1 lists identified compounds and their estimated concentration based on response factors generated from authentic standards. The data show that these waters contain several compounds which can be analyzed conveniently by the CLSA procedure. This method is best suited for the isolation and detection of semivolatile and semipolar compounds from water. However, it is of limited use for volatile compounds, such as low-molecular-weight aldehydes and ketones, and very polar compounds, such as acids.

In raw aqueduct water (Figure 1) straight-chain aldehydes (C_9-C_{11}) are commonly observed with nonanal and decanal being most abundant. In addi-

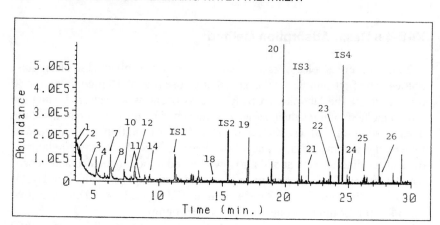

Figure 1.   Total ion chromatogram of CLSA of raw aqueduct water.

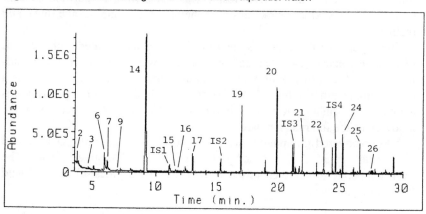

Figure 2.   Total ion chromatogram of CLSA of post-ozone-treated water.

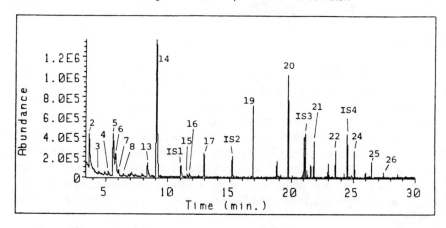

Figure 3.   Total ion chromatogram of CLSA of finished water.

Table 1. Compounds Identified in Raw Aqueduct, Post-Ozone-Treated, and Finished Water by CLSA

Code	Compound	Degree of Identification	Estimated Concentration (ng/L)		
			Raw	Post-O_3	Finished
1	Trichloroethylene	T	tr	nd	nd
2	Bromodichloromethane	T	tr	tr	tr
3	Toluene	C	tr	tr	tr
4	Carbon tetrachloride	T	tr	nd	tr
5	Dibromochloromethane	T	nd	nd	tr
6	Hexanal	C	8.1	578	401
7	Tetrachloroethylene	C	tr	tr	tr
8	Dichloroiodomethane	T	tr	nd	tr
9	Methylpentanone	T	nd	tr	nd
10	Methylcyclopentanone	T	tr	nd	nd
11	Xylenes	T	tr	nd	nd
12	Methylpentanol	T	tr	nd	nd
13	Bromoform	T	nd	nd	tr
14	Heptanal	C	7.4	2900	2000
IS1	2-Ethylhexanol				
15	Benzaldehyde	C	nd	tr	tr
16	Methylhexanal	T	nd	tr	tr
17	Octanal	C	47.0	214	209
18	2-Ethylhexanol	C	5.8	8.6	2.8
IS2	Chlorooctane				
19	Nonanal	C	139	618	464
20	Decanal	C	615	1130	1080
IS3	Chlorodecane				
21	Undecanal	C	74.6	495	401
22	Dodecanal	C	13.7	155	107
23	Geranylacetone	C	153	87.5	47.4
IS4	Chlorododecane				
24	Tridecanal	C	4.0	233	97.5
25	Tetradecanal	C	15.5	258	108
26	Pentadecanal	C	tr	tr	tr

T = tentative, matched with library spectral data; C = confirmed, matched with authentic spectra and retention data; nd = not detected; tr = detected at levels near the detection limit but not quantified.

tion, 2-ethylhexanol and geranylacetone (6,10-dimethyl-5,9-undecadien-2-one) were present. Toluene and xylene are found in small amounts. The origin of the tri- and tetrachloroethylene is unknown.

After ozonation (Figure 2) the concentration of strippable aldehydes increases, the major by-product being heptanal. Other aldehydes, up to C_{15}, are also identified. Geranylacetone seems to have diminished in abundance.

Analysis of finished water (Figure 3) shows a lower concentration of strippable compounds as compared to the ozonated sample. The major aldehydes remain at relatively high levels and five trihalomethanes are observed for the first time in significant amounts.

Laboratory Ozonation at Different Ozone Doses

The raw aqueduct water was treated with an ozone dose rate of 4.5 mg/minute for 20 minutes, after which time the rate was increased to 15 mg/minute for an additional 10 minutes. Water samples were periodically taken from the reactor and analyzed by CLSA. Laboratory ozonation of raw aqueduct water yielded strippable by-products much like those found in the field study. Figure 4 shows that the C_7-C_{11} aldehydes are formed at relatively higher concentrations at low ozone doses (1 mg/L) but that these diminish after higher ozone doses.

Total ion chromatograms obtained at low ozone doses were complicated by several unknown peaks. Mass spectra of these peaks suggested the existence of cyclic or unsaturated oxygen-containing compounds and indicate that unstable intermediates may be present at the initial stage of ozonation. These intermediates may be responsible for the mutagenicity of water treated at low ozone doses as reported by Bourbigot et al.[9] Further studies are in progress.

XAD-4/8 Resin Adsorption Method

Figure 5 shows the total ion chromatograms of the acidic extracts of ozonated and unozonated raw aqueduct water after XAD-4/8 resin accumula-

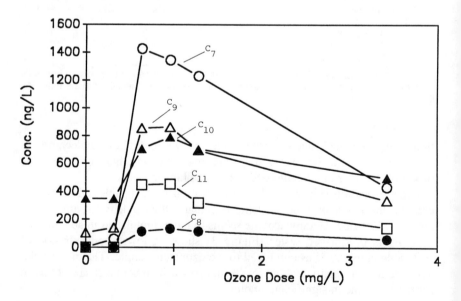

Figure 4.   Effect of ozone dose on the production of aldehydes (C_7, C_8, C_9, C_{10}, and C_{11}).

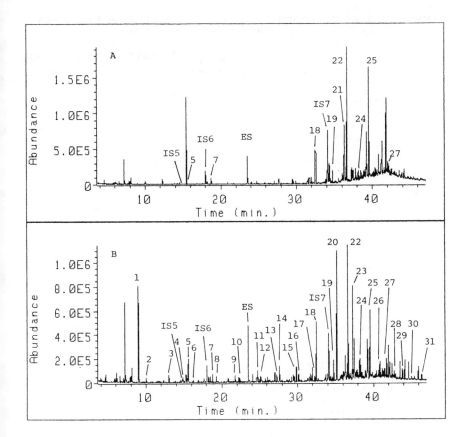

Figure 5. Total ion chromatogram of acidic extracts of raw aqueduct (A) and ozonated water (B) after XAD-4/8 resin accumulation.

tion. Each component was identified through comparison of library spectra or, when available, through mass spectra and GC retention data of authentic standards. Identified compounds are listed in Table 2 and include fatty acid methyl esters and aliphatic aldehydes. The estimated concentration of fatty acids is also included in Table 2 and shows that the concentration of most fatty acids increased after ozonation, while concentration of the naturally occurring fatty acids C_{14}, C_{16}, C_{18}, and C_{20} decreased.

Calculation of recovery efficiency for benzoic acid-d_5 through the external standard method gave an efficiency of 17.5%. This indicates fatty acid extraction efficiency to be on the order of 20%. The recovery efficiencies of other deutrated internal standards were 7% for phenol-d_5, 85% for naphthalene-d_8, and 45% for anthracene-d_{10}.

Table 2.   Compounds Identified in Unozonated and Ozonated Raw Aqueduct Water Using XAD-4/8 Resin Adsorption Method (Neutrals and Acids)

Code	Compound	Degree of Identification	Estimated Concentration (ng/L)	
			Raw	Ozonated
1	Heptanal	C		
2	Hexanoic acid, Me-ester	C	nd	27
3	Heptanoic acid, Me-ester	C	nd	970
IS5	Benzoic acid-d_5			
4	Benzoic acid, Me-ester	C	nd	tr
5	Nonanal	C		
6	Octanoic acid, Me-ester	C	nd	105
IS6	Napthalene-d_8			
7	Decanal	C		
8	Nonanoic acid, Me-ester	C	nd	210
9	Undecanal	C		
10	Decanoic acid, Me-ester	C	nd	150
ES	2-Fluorobiphenyl			
11	Dodecanal	C		
12	Undecanoic acid, Me-ester	C	nd	68
13	Tridecanal	C		
14	Dodecanoic acid, Me-ester	C	238	363
15	Tetradecanal	C		
16	Tridecanoic acid, Me-ester	C	32	170
17	Pentadecanal	T		
18	Tetradecanoic acid, Me-ester	C	1570	1175
IS7	Anthracene-d_{10}			
19	Pentadecanoic acid, Me-ester	C	434	355
20	6,10,14-Trimethylpentadecanone	T	nd	tr
21	9-Hexadecenoic acid, Me-ester	C	2165	nd
22	Hexadecanoic acid, Me-ester	C	3975	1705
23	Dibutyl phthalate	T		
24	Heptadecanoic acid, Me-ester	C	210	208
25	Octadecanoic acid, Me-ester	C	2805	765
26	Nonadecanoic acid, Me-ester	C	39	53
27	Eicosanoic acid, Me-ester	C	810	177
28	Heneicosanoic acid, Me-ester	C	nd	106
29	Docosanoic acid, Me-ester	C	144	187
30	Tricosanoic acid, Me-ester	C	nd	20
31	Tetracosanoic acid, Me-ester	C	60	140

T = tentative, matched with library spectral data; C = confirmed, matched with authentic spectra and retention data; nd = not detected; tr = detected at trace level but not quantified.

Ozonation of 9-Hexadecenoic Acid

Laboratory ozonation studies using 9-hexadecenoic acid, a compound found in raw water at relatively high levels (ca. 2 µg/L) were performed. It was predicted that heptanal, a major product found in ozonated field and laboratory samples, would be produced as a by-product of the ozone-induced decomposition of 9-hexadecenoic acid.

A one-liter aqueous solution containing 100 mg of 9-hexadecenoic acid was prepared and ozonated for two minutes at an ozone dose of 15 mg/min. A 50-mL aliquot was then acidified and extracted with 20 mL of diethylether. The extract was taken to dryness and esterified using a diazomethane-ether solution.

Figure 6 shows the total ion chromatogram of the methylated extract as well as identified by-products. Aqueous ozonation of 9-hexadecenoic acid (1) yields heptanal (2), heptanoic acid (3), 9-oxononanoic acid (4), and nonanedioic acid (5). Further studies of this model compound are in progress.

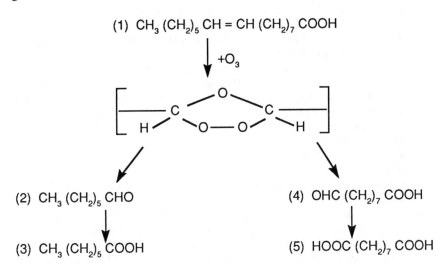

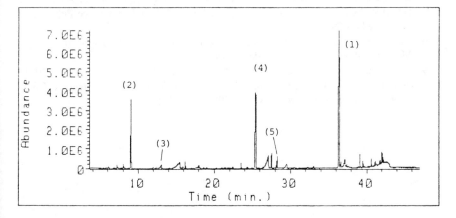

Figure 6.   Ozonation of 9-hexadecenoic acid.

ACKNOWLEDGMENTS

This work is sponsored by the Los Angeles Department of Water and Power with funds obtained in part from the U.S. EPA under cooperative agreement CR-813188 and by the University of California Water Resources Center.

REFERENCES

1. Sievers, R. E., R. M. Barkley, G. A. Eiceman, R. H. Shapiro, H. F. Walton, K. J. Kolonko, and L. R. Field. "Environmental Trace Analysis of Organics in Water by Glass Capillary Column Chromatography and Ancillary Techniques," *J. Chromatog.* 142:745–754 (1977).
2. Lawrence, J., H. Toshine, F. I. Onuska, and M. E. Comba. "The Ozonation of Natural Waters: Product Identification," *Ozone Sci. Eng.* 2:55–64 (1980).
3. Schalekamp, M. "Experiences in Switzerland with Ozone, Particularly in Connection with the Charge of Undesirable Elements Present in Water," International Ozone Association, Ozone Technology Symposium and Exposition, Los Angeles, California (1978).
4. Killops, S. D. "Volatile Ozonation Products of Aqueous Humic Material," *Water Research* 20:153–165 (1986).
5. Glaze, W. H. "Reaction Products of Ozone: A Review," *Environ. Health Pers.* 69:151–157 (1986).
6. Grob, K., and F. Zurcher. "Stripping of Trace Organic Substances from Water-Equipment and Procedure," *J. Chromatog.* 117:285–294 (1976).
7. Krasner, S. W., C. J. Hwang, and M. J. McGuire. "Development of a Closed-Loop Stripping Technique for the Analysis of Taste- and Odor-Causing Substances in Drinking Water," in *Advances in the Identification and Analysis of Organic Pollutants in Water, Vol. 2* (Michigan: Ann Arbor Science Publishers, Inc., 1981), 689–710.
8. Bader, H., and J. Hoigne. "Determination of Ozone in Water by the Indigo Method," *Water Res.* 15:449–456 (1981).
9. Bourbigot, M. M., M. C. Hascoet, Y. Levi, F. Erb, and N. Pommery. "Role of Ozone and Granular Activated Carbon in the Removal of Mutagenic Compounds," *Environ. Health Pers.* 69:159–163 (1986).

SECTION VI

Granular Activated Carbon

The Role of Granular Activated Carbon in the Reduction of Biohazards in Drinking Water

M. Wilson Tabor, Institute of Environmental Health, University of Cincinnati Medical Center, Cincinnati, Ohio

INTRODUCTION

The presence of trace levels of organic compounds in drinking water has heightened concern among the scientific community and throughout the public sector as to the human health impact of chronic exposure to these contaminants. These organic chemicals encompass a broad range of natural and man-made chemicals, some of which now occur at lower concentrations in drinking water than a decade ago due to the implementation of federal legislation like the Safe Drinking Water Act of 1974, PL 93–523, and the Clean Water Act of 1977, PL 95–217. However, in a recent evaluation of the epidemiological data on the relationship of cancer to drinking water quality, Crump and Guess[1] stated that "further studies of the identities, carcinogenicity, mutagenicity, mode of formation and practical methods of removal are needed for the organic contaminants in drinking water." As to the origin of toxic organic contaminants, the data suggest two probable sources: anthropogenic chemicals in the raw water from industrial discharges into the raw water supply, e.g., a river,[2] and conversion of nontoxic to toxic chemicals due to disinfection practices used in the preparation of drinking water.[3,4] The identities, carcinogenicity, and mutagenicity of the organic contaminants has been the focus of the research efforts of not only our laboratory,[5–9] but also several other laboratories.[10–30] These studies have elucidated the identity of several potent mutagens, including 3-chloro-4-(dichloromethyl)-5-hydroxy-2(5H)-furanone,[31] resulting from the chlorination of raw water containing humic materials and 3-(2-chloroethoxy)-1,2-dichloropropene[7] most probably of industrial origin.

Although biohazardous compounds of natural and industrial sources such as these have been identified, a more imminent threat to the public water supplies was recently brought to the forefront by the disaster that occurred in Switzerland thereby contaminating the Rhine River.[32] Those water supplies drawing on surface sources such as rivers, like the Rhine, are vulnerable to contamination due to major industrial accidents. For example, the Ohio River not only is a major source of water for publicly owned drinking water suppliers, but also is a major site for industry and attendant riverborne commercial transportation in the United States. Up river from the metropolitan area of Cincinnati, Ohio, there are 269 permitted discharge sources into the river, of which 149 are industrial. Additionally, in 1985,[33] the Ohio River was the site of transport via barge of more than 170 million tons of raw materials, of which 30 million tons were classified as hazardous. Consider that in a typical year such as 1985, 58 major barge accidents occurred[33] above Cincinnati on the river. Many of these accidents necessitated the implementation of additional drinking water treatment steps such as the use of powdered activated charcoal and/or closing of intake valves for several days to await the passage of the contaminant plumes.[34] The vulnerability of surface water to industrial accidents and spills, as well as the presence of trace levels of biohazardous compounds in finished drinking water prepared by conventional methods, has led to the proposed use of granular activated carbon in the preparation of safe drinking water.[35]

Recently, we have examined the capacity of a full-scale granular activated carbon (GAC) treatment bed designed to remove total organic carbon and Ames test mutagenicity[36,37] from chlorinated drinking water during a 35-week period. In that study, residual organics from 14 sample sets were extracted by XAD[36] and analyzed in an Ames test using strains TA98 and TA100. Each sample set consisted of: (a) settled raw water; (b) chlorinated, sand filtered, finished water that was influent to the GAC system; (c) the GAC system effluent; (d) water sample "b" stored three days to simulate distribution; (e) water sample (c) rechlorinated to 2 mg chlorine/L and stored for three days; (f) finished water as described in "b" but taken at a distant distribution point. Removal of TOC by the GAC decreased from 92% to 35% over the first 15 weeks and maintained that level throughout the remainder of the 35-week study. In previous studies (e.g., DeMarco et al.[38,39]), the GAC was considered exhausted once such a plateau had been reached. However, in our study, all mutagenic activity detected for the residual organics in sample "b," which arose as a result of the primary chlorination, was removed by GAC. Moreover, mutagenic activity was not detected from any samples of residue organics even following its rechlorination, sample "e." Thus, within the limits of our procedures, GAC treatment during the 35 weeks continually removed both mutagens and the potential for mutagen formation following rechlorination of effluent water. From that study we concluded that mutagenesis monitoring could be used as an additional test of GAC performance.

These studies raised additional questions, particularly with regard to the nature and the identity of the mutagens removed from the water by GAC. This chapter describes our examination of residue organics isolated from such GAC after extensive use for water treatment. We have examined this carbon as a convenient source of residue organics for the isolation and identification of drinking water mutagens.

EXPERIMENTAL

Instrumentation

High-performance liquid chromatography (HPLC) separations were performed on a Water Associates (Milford, MA) Model ALC/GPC 204, fitted with a model 440 fixed wavelength, 254 nm, detector and a model 441 variable wavelength detector set at 345 nm. A Waters Z-module radial compression (RCM) column unit was utilized for preparative scale, mg level, separations. RCM columns, 8 mm by 10 cm, used in separations, were packed with either 10 μm silica particles (PORASIL®), or 10 μm silica particles bonded with octadecylsilane (C18 BONDAPAK®). HPLC system operation was according to methods described previously.[5,40,41]

Gas chromatography (GC) analyses were performed on a Perkin Elmer Model 3920 (Norwick, CT) flame ionization unit fitted with a glass 6-ft-by-0.25-in. i.d. column containing 5% OV17 on 100 mesh Gas Chrom Q. Data were continuously collected and analyzed using a Hewlett-Packard (Palo Alto, CA) Model 3390A computing integrator.

Mass spectrometry (MS) analyses were performed on a Kratos MS 80 (Manchester, UK) high-performance spectrometer as previously described.[7] Data were collected continuously during MS runs and processed on a Data General (Westboro, MA) Nova/4C DS-55 data system. Computer interaction, data display, and output were via Hewlett-Packard Models 2649C graphics terminal and 9876A printer systems. Samples were introduced via GC utilizing a Carlo Erba Series 4160 GC (Milan, Italy) fitted with an SE54 30-m-by-1-mm fused silica, film thickness of 0.25 μM capillary column.

Granular Activated Carbon Samples and Other Chemicals

Both virgin (control samples) and spent GAC were obtained from the Cincinnati Water Works (Cincinnati, Ohio). The virgin GAC had been freshly regenerated according to methods described by De Marco et al.[38,39] The spent GAC had been used extensively (for more than 6 months) for the treatment of chlorinated drinking water, as described by De Marco et al.[38,39]

Type I water[42] for HPLC and for the preparation of other aqueous solutions was purified using a Continental Millipore Water Conditioning System

(El Paso, TX) as described previously.[5] HPLC-grade solvents, acetonitrile, hexane, and methylene chloride, were obtained from Fisher Scientific (Cincinnati, OH). All HPLC solvents were degassed immediately prior to use by 15 min of sonication while under reduced pressure.

4-Nitrothiophenol, from Aldrich Chemicals (Milwaukee, WI), was recrystallized according to the procedure of Barnett and Jenks,[43] and stock solutions were maintained as recommended by Cheh and Carlson.[44] Sodium sulfate, reagent grade (Fisher Scientific), was muffled at 500°C overnight prior to use. All other chemicals were of reagent grade and were used without further purification.

Mutagenicity Testing

Bioassays were performed using the *Salmonella* microsome mutagenicity assay of Ames,[45,46] who provided strains TA98 and TA100. The nitroreductase deficient strains, TA98 NA and TA100 NR, were obtained from H. Rosenkranz[47] and utilized in the *Salmonella* microsome mutagenicity assay as described by Cheh et al.[44] Characteristic properties of the bacteria were verified for each fresh stock, and their mutagenic properties were again verified using positive and negative controls as part of each experiment. Microsomal activation requiring mutagenesis tests utilized polychlorinated biphenyl mixture Aroclor 1254 induced rat liver microsomes, 9000 × g supernatant fraction (S9), from Litton Bionetics (Kensington, MD). Mutagenesis assays without (–S9) and with (+ S9) microsomal activation were conducted as previously described.[7,40] The detection of mutagenesis in experimental samples was based upon a dose-dependent response exceeding the zero dose (spontaneous control value) by at least twofold, i.e., the ratio of total revertant colonies per plate to spontaneous colonies per plate is $\geqslant 2$. In some situations, i.e., HPLC subfractions, where the amount of sample was limiting, semiquantitative determinations of mutagenesis were made as previously described.[7,40] All recoveries of bioactivity from concentrated or fractionated residue organic samples were based upon an expression of mutagenesis per liter equivalent, representative of the original water sample.

4-Nitrothiophenol Experiments

Experiments to inactivate direct-acting mutagens via reaction with the nucleophile, 4-nitrothiophenol (NTP), were conducted according to the recommendations of Cheh and Carlson.[44] Stock solutions of NTP, 10 to 20 mg/mL, were maintained at –20°C as 50:50 aqueous ethanol solutions acidified to pH 4.5. Concentrations were determined by dilution of the stock solution 1:1000 into 0.1 M sodium phosphate buffer, pH 7.4, measurement of the absorbance of this solution at 410 nm, and calculations using a molar absorptivity of 13,700 $m^{-1}cm^{-1}$. Working solutions of NTP were prepared by diluting the stock solution into $0.05\,M$ sodium phosphate buffer, pH 7.4, to a concentration range of $6.5 \times 10^{-4}M$ to $2.6 \times 10^{-2}M$.

The reaction of NTP with samples containing direct-acting TA98 and TA100 mutagens was accomplished by incubating the sample with NTP at room temperature for 30 minutes following which this mixture was bioassayed for mutagenesis using TA98 NR and TA100 NR. Ethanol solutions of 2-(chloromethyl)naphthalene (CMN) and 1,3-dichloroacetone were used as positive mutagen controls for the NTP reaction and bioassay for TA98 NR and TA100 NR mutagenesis, respectively, as described by Cheh and Carlson.[44] Equal molar amounts of the positive mutagen control and NTP were combined in $0.1\,M$ sodium phosphate buffers, pH 7.4, and allowed to react 30 minutes at room temperature. Portions of the reaction mixtures were bioassayed for mutagenicity with the TA98NR and TA100NR strains. The remaining portion of the CMN-NTP mixture was stored at 4°C for use as a standard in HPLC method development.

Gas Chromatography and Mass Spectroscopy

Gas chromatographic analyses of HPLC subfractions and SEP-PAK concentrates were performed by the slow injection of 2 to 7.5 μL of the appropriate hexane/acetone, acetonitrile/water, or methylene chloride solution. The nitrogen carrier gas was flowing at 70 mL/min, and the temperatures of the injector and detector were 200°C and 320°C, respectively. A linear temperature program from 80°C to 270°C at 7.5°C/min was initiated at the time of injection. Weight values for sample constituents were calculated based on peak areas compared to those obtained from chromatography of repeated injections of 2.0 μL of a 2.6 mg/mL methanol solution of a standard base neutral mixture.

High-resolution mass spectrometry analyses were performed in the presence of the internal mass standard perfluorokerosenes (PFK). The conditions for EI spectra were as follows: ionizing current, 1×10^{-4} A; ionizing energy, 40 eV; accelerating voltage, 4kV; scan range, 20–600 m/e; scan speed of 5.5 sec for 3000 resolution and 13 sec for 7500 resolution; scan internal, IS. Samples were introduced via direct insertion probe or capillary GC. For direct insertion inlet, 5 to 50 μL of a methylene chloride solution of sample were evaporated with a capillary sample holder which was placed in a shaft tip of the direct insertion probe. The probe was air cooled at 20°C, the air shut off, and then the temperature of the probe was increased at 10°C/min while continuously monitoring the total ion current and collecting spectral data. Capillary introduction of samples were under the following GC conditions: injection temperature, 50°C; oven temperature at 50°C for 75 sec after injection then programmed at 250°C at 10°C/min; separator temperature, 250°C; helium carrier gas at 20 mL/min with a 20-mL/min flow make-up to the separator. Five hundred nanoliters of methylene chloride or hexane/

acetone solutions of sample were introduced by the cold on-column splitless injection techniques of Grobe and Neukom.[48] After a 70-S solvent divert, the valve to the MS was opened and data were collected. Isobutane chemical ionization (CI) spectra at 1000 resolution were obtained on samples introduced by capillary GC at an ionization current of 1.5×10^{-3} A. Other GC/MS conditions were as described above.

Extraction of GAC and Preliminary Fractionation of Residue Organics

Both the virgin and spent GAC were extracted 24 hr with methylene chloride using a Soxhlet extractor, following which any traces of water were removed by passage of the extracts through anhydrous sodium sulfate. The solvent was removed via rotary evaporation and the resulting residue organics, SI, were partitioned by suspension into hexane. Insolubles were removed by centrifugation, $1500 \times g$, following which the precipitate, SII, the supernatant fluids, SIII, and the originally isolated residue organics were assayed for mutagenesis. This extraction/separation scheme is summarized in Figure 1.

HPLC Separation of Mutagenic Mixtures

Separation of SIII

A 20-G equivalent sample of SIII was injected as a hexane solution into the HPLC unit fitted with a Waters radial compression module (RCM) unit containing an 8-mm-by-10-cm column packed with 10-μm silica particles (PORASIL®). Following injection, the column was flushed with hexane at 2.0 mL/min for 2 min then the solvent was changed to an 80% hexane/20% methylene chloride solvent via a linear gradient of 2 min. This solvent was used for a 5-min isocratic elution of the column, then was changed to a mobile phase of 30% hexane/70% methylene chloride via a 3-min linear gradient. Following a 2-min linear gradient, the column was eluted for 5 min with 100% methylene chloride, then the solvent was changed to 100% methanol via a 7-min linear gradient. The column was washed with methanol until no more 254-nm absorbing material eluted. Fractions were collected (samples SIV-1 through SIV-5), gently evaporated to dryness, and the residues dissolved in DMSO for bioassay or into 0.1 M, pH 7.4 phosphate buffer for NTP inactivation experiments. An outline of the processing of samples SIII is presented in Figure 2.

Separation of NTP-Adducts

The mutagenic fractions SIV-1 and SIV-2, sensitive to inactivation by NTP, were combined; SV, inactive with NTP and the resulting NTP adducts,

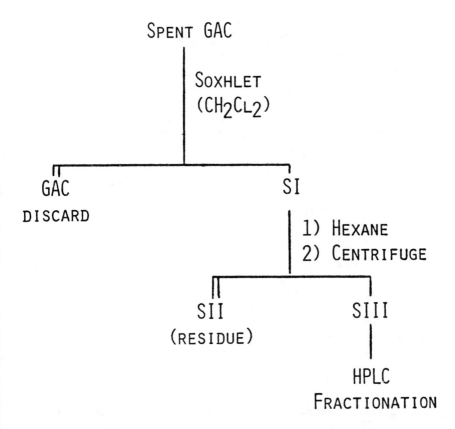

Figure 1. Extraction/separation scheme for the isolation and separation of residual organics from granular activated carbon.

were isolated via extraction of the aqueous with three equal volumes of methylene chloride. The extracts were combined and the solvent was removed by gentle evaporation. The residue was dissolved in a minimum volume, about 150 μL of acetonitrile. The solution was injected into the HPLC unit, using the RCM fitted with a preparative scale reverse phase column. The HPLC separation was accomplished via an isocratic wash with 100% water until no more 254-nm and/or 345-nm absorbing components eluted (about 5 min). Then a linear gradient of 45 min duration was initiated from 100% water to 100% acetonitrile. The gradient was followed by an isocratic wash with 100% acetonitrile until no more 254-nm and/or 345-nm absorbing components were eluted. Throughout the separation, fractions (SVII-1 through SVII-N) were collected by major (>15%) full-scale peaks. The fractions were extracted 3× with an equal volume of methylene chloride. Following reduction of the volume of the extracts to 0.2 mL, gas

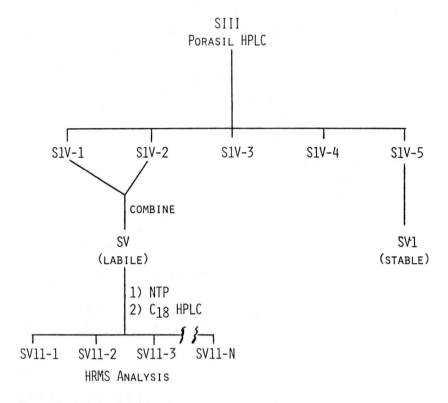

Figure 2. HPLC separations of mutagenic residual organics isolated from granular activated carbon.

chromatographic and GC/MS analyses were conducted. An outline of the processing of these samples is presented in Figure 2.

RESULTS

Isolation and Bioassay of Residue Organics from GAC

Residue organics were isolated by extraction of both the virgin and the spent GAC samples with methylene chloride. Following concentration, the extracts were bioassayed using both TA98 and TA100 (Figures 3 and 4). The extracts from the virgin GAC (open circles and diamonds) showed no mutagenic activity for either TA98 or TA100 in the absence or presence of S9. However, bioassays of the spent GAC extracts resulted in direct-acting mutagenic activity (closed circles) for both tester strains. The presence of S9 in the assay mixture reduced the level of mutagenicity (closed diamonds)

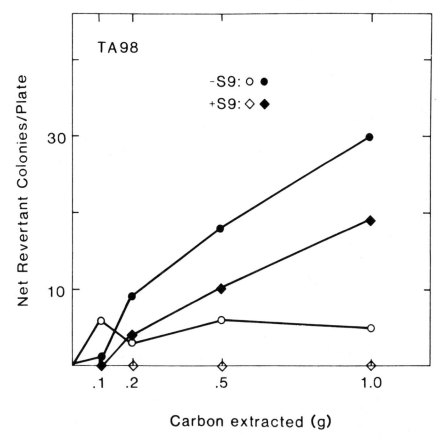

Figure 3. Mutagenic activity, to tester strain TA98, of residual organics isolated from virgin (open circles and diamonds) and spent (closed circles and diamonds) granular activated carbon in the absence (circles) and presence (diamonds) of metabolic activation (S9).

for both strains. During the evaporation of the extraction solvent, it was noted that some material precipitated, suggesting the presence of highly polar or ionic materials. Therefore, the extracts were partitioned with a nonpolar solvent, hexane, which yielded two fractions, a hexane-insoluble fraction of residue organics, SII, and a soluble fraction, SIII. Bioassay of these two fractions showed the majority of the TA100 direct-acting activity to be in the hexane solubles, SIII (Figure 5). The total mutagenic activity in the original GAC extract, SI, the hexane insoluble fraction, SII, and the soluble fraction, SIII, are summarized in Table 1. The total mutagenic activity recovered in SII and SIII was greater than the amount of activity in the original extract, SI. This indicates that antagonists were present in SI that masked the total activity of the sample. The partitioning of residual organics into hexane

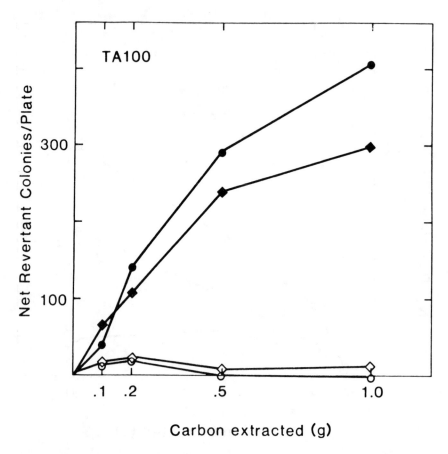

Figure 4. Mutagenic activity, to tester strain TA100, of residual organics isolated from virgin (open circles and diamonds) and spent (closed circles and diamonds) granular activated carbon in the absence (circles) and presence (diamonds) of metabolic activation (S9).

separated the antagonists from the bulk of the TA100 mutagens. The separation of antagonists to mutagenic activity for drinking water mutagens has been observed in our previous studies[5,7] and by others (e.g., Baird et al.[10-12]).

HPLC Fractionation and Studies of the Direct-Acting TA100 Mutagenic Activities

The direct-acting TA100 mutagen(s) in the GAC extract were of interest since we previously had purified[5] and identified[6] a TA100 mutagen, requiring S9 activation, from GAC extracts. Initial application of our coupled bioassay/analytical fractionation technique,[5,39] using reverse phase HPLC,

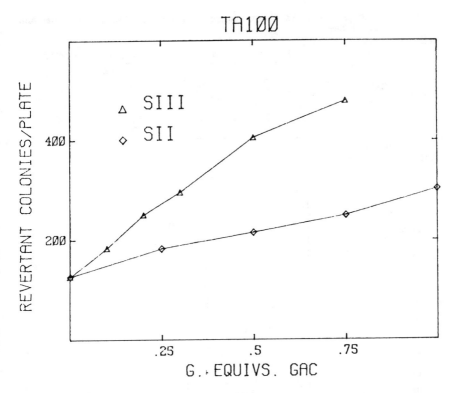

Figure 5.   Dose response of TA100 direct-acting mutagenic activity of hexane-insoluble (SII-diamonds) and hexane-soluble (SIII-triangles) residual organics isolated from spent granular activated carbon.

to the SIII residue organics resulted in the bulk (>90%) of the TA100 mutagenic activity eluting early in the polar/semipolar fractions (data not shown). This suggested that fractionation via normal phase HPLC would separate the TA100 mutagenic components. A 20-g equivalent sample of SIII was separated via normal phase HPLC using a series of isocratic and gradient solvent elutions from hexane to methylene chloride to methanol (Figure 6). Two distinct fractions of TA100 mutagenesis were isolated— fractions 1 and 5, which were termed SV and SVI, respectively. The elution conditions for these two active fractions suggested that SV contained less-polar substances than SVI.

Since the mutagenic fraction SV appeared to be semipolar, studies to further fractionate SV via reverse phase HPLC were conducted. Less than 25% of the TA100 mutagenesis was recovered from this HPLC separation. From these results, it appeared that the TA100 mutagens in SV were labile, possibly decomposing/reacting in this semipurified state. Therefore, a stability study was conducted in which SIII, SV, and SVI were stored at 4°C and

Table 1.   Mutagenesis of Residue Organics Extracted from "Spent" Granular Activated Carbon

	Net Revertant Colonies/g Equivalent GAC Extracted			
	TA9		TA100	
Sample	−S9	+S9	−S9	+S9
SI	36	19	420	341
SII	ND[a]	ND	141	NA[b]
SIII	ND	ND	590	NA

[a]ND = not detected.
[b]NA = not assayed.

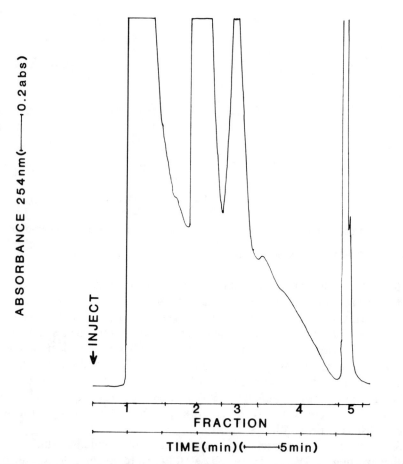

Figure 6. Normal phase high-performance liquid chromatographic separation of hexane-soluble, SIII, residual organics isolated from spent granular activated carbon. A 20-G equivalent sample of SIII was injected into the chromatography unit and fractionated, via combined isocratic and linear gradient elution, as described in the Methods section.

bioassayed over a period of 11 days. The results (Table 2) show that the TA100 mutagens in SV are lost in less than two weeks and that some component(s) of the overall matrix effect of the parent fraction, SIII, stabilized the TA100 direct-acting mutagens in this sample. Further, the TA100 mutagens in the more polar subfraction, SVI, appear to be stable.

Trapping of Direct-Acting TA100 Mutagens with NTP

The instability of the direct-acting TA100 mutagenic components of SV could possibly be due to the reactive nature of these compounds. This is not unexpected, since it is generally held that direct-acting mutagens are strong electrophiles which exhibit their mutagenic properties by reacting with nucleophilic sites of the cellular DNA. The reaction of mutagenic components with a strong nucleophile in complex mixtures of residue organics would allow for the nucleophile to stabilize these labile mutagens for further isolation/identification studies. To stabilize such mutagens, Cheh and Carlson[44] reported studies on the use of 4-nitrothiophenol (NTP) as a trapping reagent. Both SV and SVI were reacted with NTP, and the resulting mixtures were assayed for direct-acting TA100 mutagenesis using strain TA100 NR. This strain is a tested strain similar to TA100, but genetically deficient in the enzyme nitroreductase, thereby preventing the reduction of the nitro group of the reagent NTP. Such a reduction may lead to reductive intermediates, e.g., a nitroso compound, producing possible mutagenic by-products from the NTP. 1-(Chloromethyl)naphthalene (CMN), a direct-acting TA100 mutagen, was used as a control. The results of this experiment are summarized in Table 3. The direct-acting mutagenic components in SV appeared to react with NTP, whereas those in SVI appeared to be insensitive to the nucleophilic NTP.

HPLC Separation of NTP-Adducts

Results of the previous experiment showed that the direct-acting TA100 mutagens could be trapped as NTP-adducts. Since these adducts were still part of a complex mixture, an HPLC method was developed to separate the NTP-adducts for compound identification. An outline of the separation strategy for the direct-acting mutagens in sample SIV is shown in Figure 2.

The HPLC method was developed using the model reaction mixture of 2-(chloromethyl)naphthalene with NTP. Ten microliters of a solution containing 25 μg of CMN-NTP product were injected into an HPLC system fitted with a reverse phase column. A representative chromatogram of the separation of this model mixture is shown in Figure 7. Fractions were collected and the major component was identified via mass spectrometry, data not shown, as the CMN-NTP adduct. The second most prominent compo-

Table 2. Stability of Direct-Acting TA100 Mutagens in Extracts/Subfractions Isolated from Residue Organics of "Spent" Granular Activated Carbon

	Extract/Subfraction TA100 Mutagenesis (net revertant colonies, − S9, per g equiv. GAC)		
Day	SIII	SV	SVI
1	590	200	31
6	600	135	49
11	580	0	48

Table 3. Reactivity of HPLC Subfractions of "Spent" Granular Activated Carbon Residue Organics to 4-Nitrothiophenol

Dose of NTP (μg)	Per Cent Remaining Direct-Acting TA100[a] Mutagenesis for HPLC Subfraction	
	SV	SVI
0	100	100
2	84	98
5	36	94

[a]Mutagenesis assays were conducted with mutants to TA100 not containing active nitroreductase, TA100NR.

nent, eluting just prior to the CMN-NTP adduct, was identified via mass spectrometry as the disulfide reaction product of two NTP moieties.

Sample IV was reacted with NTP and separated via HPLC in the presence and absence of an added CMN-NTP standard (chromatograms in Figures 8 and 9, respectively). Fractions were collected for analysis via mass spectrometry. The presence of the CMN-NTP adduct was verified in fraction 13 (Figure 8), and the presence of the NTP-disulfide dimer was found in fraction 14 (Figures 8 and 9) of both samples. Although efforts to identify the other NTP-adducts in these fractions via mass spectrometry are in progress, only one adduct of NTP, fraction 17, has been tentatively identified as a N-(2,6-diethylphenyl)-N-methoxymethylacetamide-NTP adduct. This compound appears to be related to the herbicide Alachlor but positive identification will require further structural elucidation studies. Furthermore, this adduct was estimated to represent about 2% of the total NTP adducts in SV.

DISCUSSION

These studies and our previous investigations of GAC[36,37] support the recommendation for the use of GAC in the removal of waterborne mutagens. The results showed that the treatment of water with GAC efficiently and preferentially removed mutagens relative to other chemicals that made up the water TOC, and this removal was shown to continue long after the percent of total TOC reduction reached a steady state. The nature and the identity of

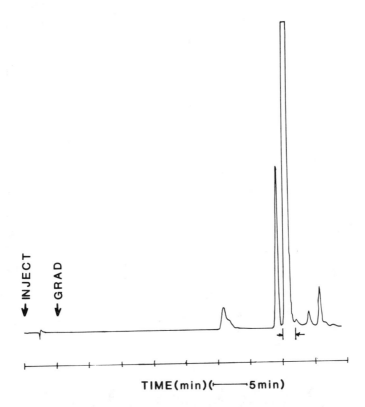

Figure 7. Reverse phase high-performance liquid chromatographic separation of 25 μg of the 2-(chloromethyl)naphthalene-nitrothiophenol reaction product. The sample was injected into the chromatographic unit and fractionated via combined isocratic and linear gradient elution as described in the Methods section.

the mutagens removed from the water by GAC was the focus of the study reported herein. To examine the residue organics, methods were developed to isolate the residues from the GAC and to separate this complex mixture of residue organics into distinct groups of mutagens for characterization. The most significant result was the isolation of two distinct fractions of TA100 direct-acting mutagens from this complex mixture via HPLC, one of which was labile in terms of loss of bioactivity whereas the other was stable. This led to the application of a previously reported method[44] to the stabilization of the labile mutagens as NTP-adducts. The adducts were separated, and one was tentatively identified as a N-(2,6-diethylphenyl)-N-methoxymethylacetamide-NTP adduct. The significance of this finding is unknown at this time in that the adduct would indicate the presence of a derivative related to Alachlor®, a herbicide previously reported not to be mutagenic in *Salmonella* bioassays.[49,50] Furthermore, this adduct was estimated to represent less than 2% of the total NTP-adducts in the labile fraction of mutagens.

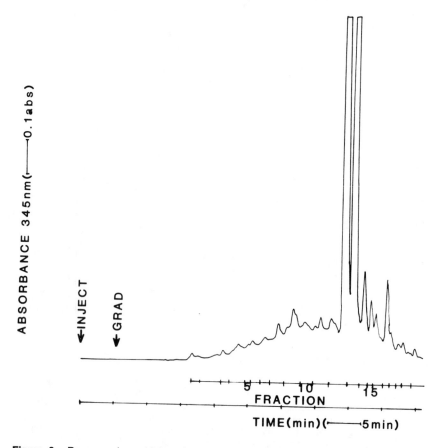

Figure 8. Reverse phase high-performance liquid chromatographic separation of the NTP-adduct products from the reaction of nitrothiophenol and the labile-mutagen-containing fraction of residue organics, sample SIV, in the presence of the CMN-NTP adduct standard, Figure 7. The sample was injected into the chromatographic unit and fractionated via combined isocratic and linear gradient elutions as described in the Methods section.

However, a more general conclusion can be drawn from these results in that the approach, used in this report, can be used for the identification of labile mutagens isolated from complex mixtures of residual organics in environmental samples. This was the intent of Cheh and Carlson[44] where they introduced the NTP-adduct method, although they did not isolate any NTP-adducts for compound identification.

The success in the isolation of mutagens from the GAC further supports our previous studies[36,37] in showing that GAC effluent drinking water was mutagen-free, whereas the influent waters contained measurable levels of mutagenic residual organics. Therefore, these results support the recommen-

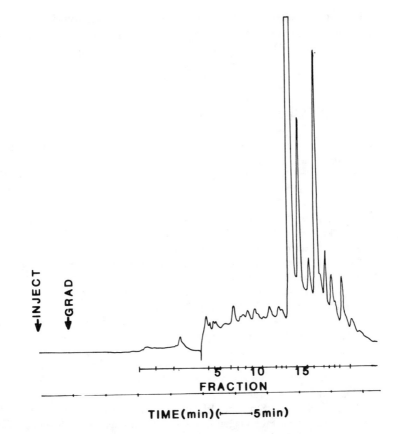

Figure 9. Reverse phase high-performance liquid chromatographic separation of the NTP-adduct products from the reaction of nitrothiophenol and the labile-mutagen-containing fraction of residue organics, sample SIV, in the absence of the CMN-NTP adduct standard, Figure 7. The sample was injected into the chromatographic unit and fractionated via combined isocratic and linear gradient elutions as described in the Methods section.

dation[35] that GAC is a viable method for removing biohazardous organics from water.

ACKNOWLEDGMENTS

Technical assistance by Ms. L. Rosenblum and mass spectrometry operation by Dr. K. Jayasimhulu are gratefully acknowledged. Appreciation is extended to Mr. R. Hutchinson, Dr. H. Kagen, and Ms. D. Gorman for valuable advice and assistance in the preparation of this chapter. This research was supported in part by grants from the USPS-NIEHS (ES00159),

USEPA (CR808603) and the National Science Foundation (PCM-82 19912). This chapter has not been subjected to U.S.EPA review. Therefore, it does not reflect the views of the Agency, and no official endorsement should be inferred.

REFERENCES

1. Crump, K. S., and H. A. Guess. "Drinking Water and Cancer: Review of Recent Findings and Assessment of Risks," Science Research Systems, Inc., prepared for Council on Environmental Quality. Springfield, VA: National Technical Information Service, U.S. Department of Commerce, Contract No. EQ10AC018 (1980), xiii.
2. Rosen, A. A. "The Foundations of Organic Pollutant Analysis," in *Identification and Analysis of Organic Pollutants in Water,* L. H. Keith, Ed. (Ann Arbor, MI: Ann Arbor Science Publishers, Inc., 1976), 3–14.
3. "Drinking Water Disinfectants," in *Environmental Health Perspectives,* F. C. Kopfler, Ed. Research Triangle Park, NC: Department of Health and Human Services. Publication No. NIH 86–218, National Institute of Environmental Health Sciences, Vol. 69 (1986), 312 pp.
4. "The Disinfection of Drinking Water," in *Drinking Water and Health, Vol. 2,* National Academy of Sciences Safe Drinking Water Committee, J. Doull, Chairman (Washington, DC: National Academy of Sciences, 1980), 5–249.
5. Tabor, M. W., and J. C. Loper. "Separation of Mutagens from Drinking Water Using Coupled Bioassay/Analytical Fractionation," *Int. J. Environ. Anal. Chem.* 8:197–215 (1980).
6. Tabor, M. W. "Structure Elucidation of 3-(2-chloroethoxy)-1,2-Dichloropropene, a New Promutagen from an Old Drinking Water Residue," *Environ. Sci. Technol.* 17(6):324–328 (1983).
7. Tabor, M. W., and J. C. Loper. "Analytical Isolation, Separation, and Identification of Mutagens from Nonvolatile Organics of Drinking Water," *Int. J. Environ. Anal. Chem.* 19:281–318 (1985).
8. Tabor, M. W. "Drinking Water: Mutagenicity and Sample Preparation Protocols," in *Guidelines for Preparing Environmental and Waste Samples for Mutagenicity (Ames) Testing: Interim Procedure,* L. R. Williams, Technical Monitor, U.S. EPA, Office of Research and Development, USEPA 600/4–85/058 (1985), 67–108.
9. Loper, J. C. "Mutagenic Effects of Organic Compounds in Drinking Water," *Mutation Res.* 76:241–268 (1980).
10. Baird, R., J. Gute, C. Jacks, et al. "Health Effects of Water Reuse: A Combination of Toxicological and Chemical Methods for Assessment," in *Water Chlorination: Environmental Impact and Health Effects, Vol. 3,* R. L. Jolley, W. A. Brungs, and R. B. Cumming, Eds. (Ann Arbor, MI: Ann Arbor Science Publishers, Inc., 1980), 925–935.
11. Baird, R. B., C. A. Jacks, and R. L. Jenkins. "A High-Performance Macroporous Resin Concentration System for Trace Organic Residues in Water," in

Chemistry in Water Reuse, Vol. 2, R. L. Jolley, et al., Eds. (Ann Arbor, MI: Ann Arbor Science Publishers, Inc., 1981), 149–169.

12. Jenkins, R. L., C. A. Jacks, R. B. Baird, et al. "Mutagenicity and Organic Solute Recovery from Water with a High-Volume Resin Concentration," *Water Res.* 17(11):1569–1574 (1983).

13. Kool, H. J., C. F. Van Kreijl, H. J. Van Kranen, et al. "The Use of XAD-Resins for the Detection of Mutagenic Activity in Water, II. Studies with Drinking Water," *Chemosphere* 10:99–109 (1981).

14. Kool, H. J., C. F. Van Kreijl, H. J. Van Kranen, et al. "Toxicity Assessment of Organic Compounds in Drinking Water in the Netherlands," *Sci. Total Environ.* 18:135–153 (1981).

15. Kool, H. J., C. F. Van Kreijl, and B. C. J. Zoeteman. "Toxicology Assessment of Organic Compounds in Drinking Water," *CRC Crit. Rev. Environ. Control* 12(4):307–359 (1982).

16. Kool, H. J., C. F. Van Kreijl, and H. Van Oers. "Mutagenic Activity in Drinking Water in the Netherlands. A Survey and a Correlation Study," *Toxicol. Environ. Chem.* 7:111–129 (1984).

17. Kool, H. J. "Health Risk in Relation to Drinking Water Treatment," in *Biohazards of Drinking Water Treatment,* (Chelsea, MI: Lewis Publishers, Inc., 1988), Chapter 1, this book.

18. Zoeteman, B. D. G., J. Hrubec, E. deGreed, and H. J. Kool. "Mutagenic Activity Associated with By-Products of Drinking Water Disinfection by Chlorine, Chlorine Dioxide, Ozone, and UV-Irradiation," *Environ. Health Pers.* 46:197–205 (1982).

19. Nestman, E. R., G. L. LeBel, D. T. Williams, and D. J. Kowbel."Mutagenicity of Organic Extracts from Canadian Drinking Water in the *Salmonella*/Mammalian Microsome Assay," *Environ. Mutagen.* 1:337–345 (1979).

20. Douglas, G. R., E. R. Nestmann, and G. Lebel. "Contribution of Chlorination to the Mutagenic Activity of Drinking Water Extracts in *Salmonella* and Chinese Hamster Ovary Cells," *Environ. Health Pers.* 69:81–87 (1986).

21. Williams, D. T., E. R. Nestmann, G. L. LeBell, F. M. Benoit, and R. Otson. "Determination of Mutagenic Potential and Organic Contaminants of Great Lakes Drinking Water," *Chemosphere* 11(3): 263–276 (1982).

22. Noordsij, A., J. Van Beveren, and A. Brandt. "Isolation of Organic Compounds from Water for Chemical Analysis and Toxicological Testing," *Intern. J. Environ. Anal. Chem.* 13:205–217 (1983).

23. Van Der Gaag, M. S., A. Noordsij, C. M. Poels, and J. C. Schippers. "Orienterend onderzoek met analytisch-chemische en genotoxicologische meetmethoden near het effect rian water-behandelingsprocessen (in Dutch)," *H₂O*15:539–558 (1982).

24. Wilcox, P., and S. Williamson. "Mutagenic Activity of Concentrated Drinking Water Samples," *Environ. Health Pers.* 69:141–149 (1986).

25. Wilcox, P., and S. Denny. "Effect of Dechlorinating Agents on the Mutagenic Activity of Chlorinated Water Samples," in *Water Chlorination: Environmental Impact and Health Effects, Vol. 5,* R. L. Jolley, R. J. Bull, W. P. Davis, S. Katz, M. H. Roberts, Jr. , and V. A. Jacobs, Eds. (Chelsea, MI: Lewis Publishers, Inc., 1985), 1329–1339.

26. Meier, J. R., R. D. Lingg, and R. J. Bull. "Formation of Mutagens Following Chlorination of Humic Acid: A Model for Mutagen Formation During Drinking Water Treatment," *Mutation Res.* 118:25–41 (1983).
27. Meier, J. R., H. P. Ringhand, W. E. Coleman, J. W. Munch, R. P. Streicher, W. H. Kaylor, and K. M. Schenck. "Identification of Mutagenic Compounds Formed During Chlorination of Humic Acid," *Mutat. Res.* 157:111–122 (1985).
28. Meier, J. R., H. P. Ringhand, W. E. Coleman, K. M. Schenck, W. E. Munch, J. W. Streicker, R. P. Kaylor, and F. C. Kopfler. "Mutagenic By-Products from Chlorination of Humic Acid," *Environ. Health Pers.* 69:101–107 (1986).
29. Cognet, L., Y. Courtois, and J. Mallevialle. "Mutagenic Activity of Disinfection By-Products," *Environ. Health Pers.* 69:165–175 (1986).
30. Bourbigot, M. M., M. C. Hascoet, Y. Levi, F. Erb, and N. Pommery. "Role of Ozone and Granular Activated Carbon in the Removal of Mutagenic Compounds," *Environ. Health Pers.* 69:159–163 (1986).
31. Meier, J. R., R. B. Knohl, W. E. Coleman, H. P. Ringhand, J. W. Munch, W. H. Kaylor, R. P. Streicher, and F. C. Kopfler. "Studies on the Potent Bacterial Mutagen, 3-Chloro-4-(dichloromethyl)-5-hydroxy-2(5H)-furanone: Aqueous Stability, XAD Recovery and Analytical Determination in Drinking Water and in Chlorinated Humic Acid Solutions," *Mutation Res.* (in press).
32. P. L. Layman, "Rhine Spills Force Rethinking of Potential for Chemical Pollution," *Chem. Eng. News* 65(8):7–11 (1987).
33. "1985 Annual Report," Ohio River Sanitation Commission, Dixie Terminal Building, Cincinnati, Ohio.
34. Hartman, D., Chief Chemist, Cincinnati Water Works, 5651 Kellogg Avenue, Cincinnati, OH. Personal communication (22 June 1987).
35. "An Evaluation of Activated Carbon for Drinking Water Treatment," in *Drinking Water and Health, Vol. 2,* National Academy of Sciences Safe Drinking Water Committee, J. Doull, Chairman (Washington, DC: National Academy of Sciences, 1980), 251–380.
36. Loper, J. C., M. W. Tabor, L. Rosenblum, and J. DeMarco. "Continuous Removal of Both Mutagens and Mutagen Forming Potential by a Full Scale Granular Activated Carbon Treatment System," *Environ. Sci. Technol.* 19:333–339 (1985).
37. Loper, J. C., M. W. Tabor, L. Rosenblum, and J. DeMarco. "Mutagens of Chlorinated Drinking Water Removed by Treatment with Granular Activated Carbon," in *Water Chlorination: Environmental Impact and Health Effects, Vol. 5,* R. L. Jolley, R. J. Bull, W. P. Davis, S. Katz, M. H. Roberts, Jr., and V. A. Jacobs, Eds. (Chelsea, MI: Lewis Publishers, Inc., 1985), 1329–1339.
38. DeMarco, J., A. A. Stevens, and D. J. Hartman. "Application of Organic Analysis for Evaluation of Granular Activated Carbon Performance in Drinking Water Treatment," in *Advances in the Identification and Analysis of Organic Pollutants in Water,* L. H. Keith, Ed. (Ann Arbor, MI: Ann Arbor Science Publishers, Inc.), 907–940.
39. De Marco, J., R. Miller, D. Davis, and C. Cole. "Experiences in Operating a Full Scale Granular Activated Carbon System with On-Site Reactivation," in *Treatment of Water by Granular Activated Carbon,* M. J. McGuire and I. H. Suffet, Eds. Advances in Chemistry Series, Vol. 202 (Washington, DC: American Chemical Society, 1983), 525–563.

40. Tabor, M. W., and J. C. Loper. "Mutagen Isolation Methods: Fractionation of Nonvolatile Residue Organics from Aqueous Environmental Samples," *Adv. Chem.* 214:401–424 (1987).

41. Tabor, M. W. "High Performance Liquid Chromatography: Mutagenicity Sample Preparation Protocols," in *Guidelines for Preparing Environmental and Waste Samples for Mutagenicity (Ames) Testing: Interim Procedures,* L. R. Williams, Technical Monitor, U.S. EPA, Office of Research and Development, USEPA/600/4–85/058 (1985), 227–255, 1985.

42. "IERL-RTP Procedures Manual: Level 1 Environmental Assessment," U.S. EPA, Office of Research and Development, USEPA/600/7–78–201 (July 1978).

43. Barnett, R. E., and W. P. Jencks. "Diffusion-Controlled and Concerted Base Catalysis in the Decomposition of Hemithioacetals," *J. Am. Chem. Soc.* 91:6758 (1969).

44. Cheh, A. M., and R. E. Carlson. "Determination of Potentially Mutagenic and Carcinogenic Electrophiles in Environmental Samples," *Anal. Chem.* 53:1001 (1981).

45. Ames, B. N., J. McCann, and E. Yamasaki. "Methods for Detecting Carcinogens and Mutagens with the *Salmonella*/Mammalian-Microsome Mutagenicity Test," *Mutation Res.* 31:347 (1975).

46. Williams, L. R., and J. E. Preston. "Interim Procedures for Conducting the *Salmonella*/Microsomal Assay (Ames Test)," U.S.EPA, Office of Research and Development, EMSL-LV, Las Vegas, NV, USEPA/600/4–82–068 (March 1983).

47. Rosenkranz, H. S., and W. T. Speck. "Mutagenicity of Metronidazole: Activation by Mammalian Liver Microsomes," *Biochem. Biophys. Res. Commun.* 66:520 (1975).

48. Grobe, K., Jr., and H. P. Neukom. "Factors Affecting the Accuracy and Precision of Cold On-Column Injections in Capillary Gas Chromatography," *J. Chromatog.* 189:109 (1980).

49. Moriya, M., T. Ohta, K. Watanabe, T. Miyazawa, K. Kato, and Y. Shirasu. "Further Mutagenicity Studies on Pesticides in Bacterial Reversion Assay Systems," *Mutation Res.* 116:185–216 (1983).

50. Probst, G. S., R. E. McMahon, L. E. Hill, C. Z. Thompson, J. K. Epp, and S. B. Neal. "Chemically Induced Unscheduled DNA Synthesis in Primary Hepatocyte Cultures: A Comparison with Bacterial Mutagenicity Using 218 Compounds," *Environ. Mutagen.* 3:11–32 (1981).

CHAPTER 18

Oxidation of Phenol on Granular Activated Carbon

Lina S. Chin, Richard A. Larson,* and Vernon L. Snoeyink, Institute for Environmental Studies and Department of Civil Engineering, University of Illinois at Urbana-Champaign, Urbana, Illinois

INTRODUCTION

In drinking water treatment, granular activated carbon (GAC) is widely used to remove hydrophobic organic substances because of its high adsorptive capacity. However, many inorganic[1] and organic[2,3] adsorbates in aqueous solution are catalytically oxidized by activated carbon. It has been suggested that oxygen-containing surface groups take part in these reactions.

Phenolic compounds, common contaminants in water, are quite sensitive to one-electron oxidants to form phenoxy radicals. The coupling of phenoxy radicals leads to the formation of dimers[4,5] or polymers[5] that are potentially hazardous. However, the final products of phenol oxidation are significantly dependent on the substituents of phenol, the type of catalyst, and the reaction conditions employed. In previous studies[6,7] in our laboratory, we have shown that GAC, when treated with disinfectants, becomes active in promoting such coupling reactions. We now report that several types of GAC oxidize adsorbed phenol without any applied oxidant.

EXPERIMENTAL METHODS

All GAC experiments were conducted in a fixed bed, dynamic adsorption system at room temperature. The GAC column was prepared by packing 2

---

*To whom correspondence should be addressed at the Institute for Environmental Studies, 1005 Western Ave., Urbana, Illinois 61801.

grams of virgin GAC into a glass tube. The inner diameter of the column was 1 cm and the length 8 cm. Before use the carbons were washed in deionized water several times to remove water-soluble ash and fines, and baked at 175°C for one week. The carbons were kept continuously at 103°C and cooled to room temperature in a desiccator before packing into the column.

Phenol solutions, freshly prepared in deionized-distilled water and buffered with $0.005 M$ phosphate mixture to desired pH value, were continuously pumped onto the GAC column at a flow rate of 10 mL/min. After pumping, the GAC was extracted with methylene chloride for 24 hours. The GAC extract was dried over anhydrous sodium sulfate, concentrated to a desired volume, and then analyzed by gas chromatography (GC) and gas chromatography-mass spectrometry (GC-MS) qualitatively without any derivation. Both gas chromatographs were equipped with DB-1 30M fused silica capillary columns (J. and W. Scientific, Inc., Folsom, CA). GC and GC-MS samples were temperature-programmed from 40–240°C at 4°C/min. The compounds in the sample were identified by comparing their mass spectra with literature, or authentic standard spectra. The quantitative analyses were performed by flame ionization detection (GC, splitless injector system) by comparing peak area ratios with an internal standard (anthracene, 40 ppm).

The phenol adsorption was determined by measuring the phenol concentration in the column effluent and mass balance on the whole column. The column effluents, taken periodically, were acidified with dilute HCl solution to pH less than 6 and their phenol concentrations were measured by ultraviolet/visible spectrophotometry at 270 nm. A double-beam spectrophotometer (Perkin-Elmer Lambda 3) was employed using quartz cells.

RESULTS AND DISCUSSION

Phenol-GAC Reaction

Carbons compared included F400 GAC (Calgon Corp.) with different dates of manufacture, BPL (Calgon Corp.), and NUCHAR WV-G (Westvaco). The phenol solution was applied in relatively high concentrations (30 mg/L, 10 liters) to obtain good yields of products. After phenol was adsorbed onto GAC surface, three phenolic products (Figure 1), 2,2'-dihydroxybiphenyl (C—C dimer), 4-hydroxydiphenyl ether (C—O dimer), and an unknown compound (mw 276) were found for almost all the carbons tested. Since only 2,2'-dihydroxybiphenyl was commercially available, its calibration curve (anthracene as an internal standard) was also used for quantification of 4-hydroxydiphenyl ether and the unknown product. The use of this calibration curve was valid because 4-hydroxydiphenyl ether had a close GC retention time to 2,2'-dihydroxybiphenyl and the unknown was a minor

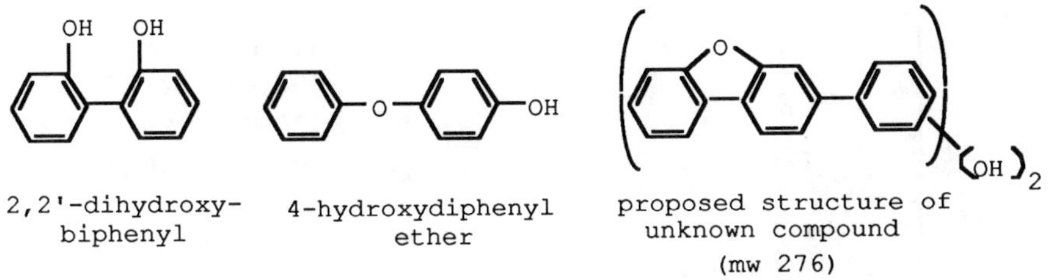

2,2'-dihydroxy-biphenyl

4-hydroxydiphenyl ether

proposed structure of unknown compound (mw 276)

Figure 1.   Identified and proposed product structures.

product. The corrections of extraction efficiencies of product yields were also limited by their commercial availabilities. The minimum amount of each product was therefore reported. Figure 2 shows the yields of the products obtained using different carbons.

Carbons of different batches showed the ability to oxidize adsorbed phenol to form hydroxylated dimers despite different activities. The total yield based on adsorbed phenol (237 mg/2 g GAC) was 6.3% in the case of F400 GAC (U.S. standard mesh 40 × 50) as shown in Figure 2. This carbon was used in the remaining experiments. Based on the phenol coupling mechanism,[4,5] phenoxy free radicals are essential intermediates to form dimers. It is most probable that phenol reacted with active sites on the GAC surface and generated phenoxy radicals for further dimerization. However, different results were observed by Voudrias[8] under similar experimental conditions in previous studies. The virgin carbon he used showed less activity toward adsorbed phenol and no phenolic dimer was detected without surface activation by free or combined chlorine. Since GAC of the same batch was no longer available, we could not confirm these findings.

Effect of pH and Phenol Loading

The yield of products was strongly dependent upon the reaction pH and the condition of phenol loading. Investigating the reactions at pH 11, 8, and 6 (Figure 3), it was observed that higher pH led to higher yields and preferential formation of C—C dimer. Raising the pH from 8 to 11, phenol adsorption decreased from 237 to 78 mg/2 g GAC; however, the yield (based on phenol adsorbed) increased from 6.3% to 15.0%. The poor adsorption of phenol at pH 11 resulted in smaller amount of total products than at pH 8. The molar ratios of C—C dimer to C—O dimer were 8.5, 3.4, and 0.86 at pH 11, 8, and 6, respectively. The formation of the phenolate anion could be important in the above observation (pKa of phenol is 9.9 in aqueous solution at 25°C). Very little is known about modification of GAC surfaces by varying pH.

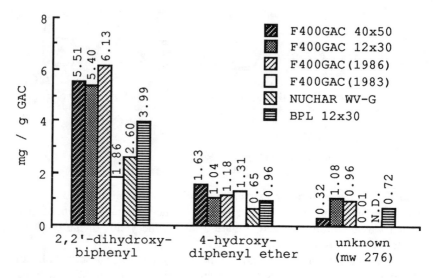

Figure 2.   Phenol-GAC reactions at pH 8.

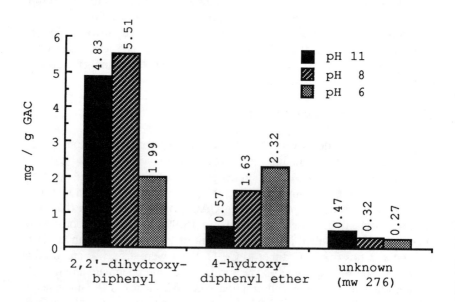

Figure 3.   Effects of pH on phenol-F400 GAC reactions.

Lowering phenol loading by changing the initial concentration (5–30 mg/L, 10 liters) of the phenol solution showed a dramatic decrease of each product (Figure 4). Total product yield versus phenol adsorption gave a linear relationship as plotted in Figure 5. This result indicates the reaction was dependent on surface coverage, which varied directly with phenol adsorption.

Biodegradation of phenol was not of concern in those experiments with lower phenol concentration, since the carbons were cleaned and baked before use, and also the running time of each experiment was only 16.6 hours, too short to have considerable biological activity without addition of inoculum and nitrogen source.

Effects of Fulvic Acid

Humic substances are representative of organic carbon in natural water. Fulvic acid extracted from soil, which has characteristics similar to those of fulvic acid in natural water,[9] was applied onto a GAC column by dissolving it in the phenol solution at two different concentrations, 7.8 and 33.3 mg/L TOC. The results shown in Figure 6 indicate that fulvic acid has two significant effects on the phenol-GAC reaction.

The presence of fulvic acid markedly favored C—O coupling relative to C—C coupling. The plot of the molar ratio of C—O to C—C dimer vs fulvic acid concentration is shown in Figure 7. Similar phenomena are noticed in

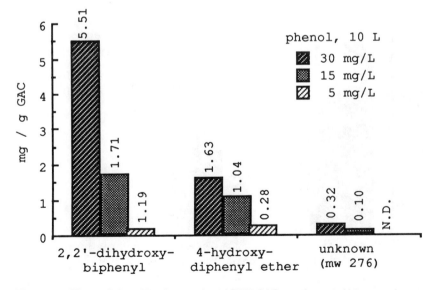

Figure 4. Effects of phenol loading on phenol-F400 GAC reactions at pH 8.

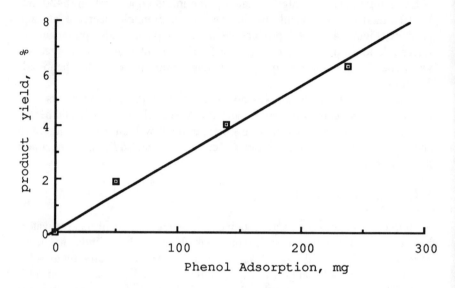

Figure 5.   Product yield of phenol-F400 GAC reactions at pH 8.

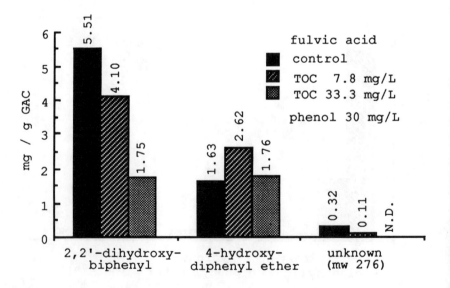

Figure 6.   Effects of fulvic acid on phenol-F400 GAC reactions at pH 8.

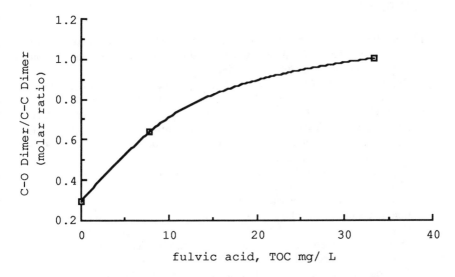

Figure 7. Effects of fulvic acid on the formation of C—O and C—C dimers in phenol-F400 GAC reactions at pH 8.

2,6-dimethylphenol oxidative coupling catalyzed by copper(I) chloride and pyridine[10]; C—O coupling is promoted more than C—C coupling by an increase of the concentration of pyridine or the ligand ratio (N/Cu ratio). Analogously, humic substances (which contain 1–6% nitrogen) are able to form complexes with metal ions.[11] Furthermore, fulvic acid inhibited the formation of all products as its concentration increased. The inhibition of fulvic acid might result from the competition with phenol for the active sites on the GAC surface or the reaction with the produced phenoxy free radicals. However, more studies are necessary to better understand the roles fulvic acid plays in the phenol-GAC reaction.

SUMMARY

Granular activated carbon was able to promote the oxidation of phenol adsorbed from aqueous solution. 2,2'-Dihydroxybiphenyl and 4-hydroxydiphenyl ether were the predominant products. The product yield was significantly dependent on the reaction pH and the phenol adsorption. The presence of fulvic acid inhibited the formation of phenolic dimers and favored C—O dimer production.

REFERENCES

1. Puri, B. R., L. R. Sharma, and D. D. Singh. "Studies in Catalytic Reactions of Charcoal. Part I. Catalytic Oxidation and Decomposition of Salt Solutions," *J. Ind. Chem. Soc.* 35:765–769 (1958).
2. Smith, E. M., S. Affrossman, and J. M. Courtney. "The Catalytic Oxidation of Creatinine by Activated Carbon," *Carbon* 17:149–152 (1979).
3. Ishizaki, C., and J. T. Cookson, Jr. "Influence of Surface Oxides on Adsorption and Catalysis with Activated Carbon," in *Chemistry of Water Supply, Treatment, and Distribution*, A. J. Rubin, Ed. (Ann Arbor, MI: Ann Arbor Science Publishers Inc., 1974), 201–231.
4. Taylor, W. I., and A. R. Battersby. *Oxidative Coupling of Phenols* (New York: Marcel Dekker, Inc., 1967).
5. Hay, A. S., H. S. Blanchard, G. F. Endres, and J. W. Eustance. "Polymerization by Oxidative Coupling," *J. Am. Chem. Soc.* 81:6335–6336 (1959).
6. Chen, A. S. C., R. A. Larson, and V. L. Snoeyink. "Importance of Surface Free Radicals in an Aqueous Chlorination Reaction (Indan-ClO_2) Promoted by Granular Activated Carbon," *Carbon* 22:63–75 (1984).
7. Voudrias, E. A., R. A. Larson, and V. L. Snoeyink. "Importance of Surface Free Radicals in the Reactivity of Granular Activated Carbon Under Water Treatment Conditions," *Carbon* 25:503–515 (1987).
8. Voudrias, E. A. "Effects of Activated Carbon on Reactions of Free and Combined Chlorine with Phenols," PhD thesis, University of Illinois, Urbana, IL (1984).
9. Weber, J. H., and S. A. Wilson. "The Isolation and Characterization of Fulvic Acid and Humic Acid from River Water," *Water Res.* 9:1079–1084 (1974).
10. Endres, G. F., A. S. Hay, and J. W. Eustance. "Polymerization by Oxidative Coupling. V. Catalytic Specificity in the Copper-Amine-Catalyzed Oxidation of 2,6-Dimethylphenol," *J. Org. Chem.* 28:1300–1304 (1963).
11. Schnitzer, M., and S. U. Khan. *Humic Substances in the Environment* (New York: Marcel Dekker, Inc., 1972).

CHAPTER 19

Microbiology of Granular Activated Carbon Used in the Treatment of Drinking Water

Gordon A. McFeters, Anne K. Camper, David G. Davies, Susan C. Broadaway, and Mark W. LeChevallier, Department of Microbiology, Montana State University, Bozeman, Montana

INTRODUCTION

Activated carbon, in the form of powdered activated carbon (PAC) and granular activated carbon (GAC), is widely used to control taste and odor problems arising from a variety of sources. Activated carbon is also useful in reducing the level of potentially harmful organic compounds in drinking water such as petroleum products, phenol, and trihalomethanes[1,2] For that reason and because of deteriorating source water quality, amendments to the National Interim Primary Drinking Water Regulations[3,4] have encouraged the utilization of activated carbon in the treatment of potable water. These observations indicate that the use of activated carbon will continue to increase in the treatment of drinking water.

The increasing use of treatment practices utilizing activated carbon has prompted interest in the microbiology of this material, since the same properties that make it an ideal medium for the adsorption of a wide range of chemicals from drinking water also promote the colonization and growth of many types of bacteria on GAC and PAC. The high degree of transition porosity and large surface area of activated carbon favor the concentration of bacterial nutrients from drinking water onto the carbon surface,[5-9] where high numbers of bacteria have been observed.[5,8,9] This sequence of events is supported by a growing body of information concerning the behavior of bacteria and their nutrients at solid-liquid interfaces[10,11] and there is some indirect evidence of bacterial growth on activated carbon surfaces in drinking water environments.[2,5-9] However, bacteria on the surface of GAC particles within a drinking water treatment system would have to be mobilized in some way to cause problems within the distribution system.

Bacteria from the filter can be transported into distribution water when particles pass the treatment barrier. The penetration of small particles of activated carbon has been demonstrated in the routine operation of filtration plants.[12-14] Such particles include fragments of GAC broken from the filter media during backwashing. In addition, bacteria may be sloughed or sheared from filter beds by hydraulic forces and enter the treated water.[15] It is possible that these mechanisms are responsible for the inoculation of distribution water, where growth might occur in sediments or other enriched environments. Such an observation has been made when bacterial regrowth of a GAC-treated and chlorinated effluent resulted in over 1000 CFU/mL in three days of incubation at 20°C.[8] However, this scenario is somewhat elusive, since it has been demonstrated that elevated populations of injured coliforms penetrated properly operated treatment barriers and were not detected on commonly used microbiological enumeration media.[16]

The effect of GAC colonization on the bacterial content of treated drinking water has been investigated by several workers. The results of such studies should provide conclusive evidence indicating whether the bacterial growth on GAC filter beds penetrated treatment barriers and contaminated treated water. A review of studies examining this question[2] indicated that the findings have been inconclusive. The data from many of the reports indicated a negligible problem, while others have shown relatively high populations of bacteria after filtration. This lack of consensus is not surprising in view of the methodological difficulties inherent in the enumeration of bacterial populations that may be attached to particulates or aggregated in drinking water. For instance, Ridgway and Olson[17] reported that 17% of the particles they examined in drinking water were colonized by average populations of 10 to 100 bacteria per particle. Accepted enumeration methods may have underestimated the actual number of bacteria in that study by 1500- to 15,000-fold. The use of different media and incubation combinations has also lead to highly variable results.[18] In addition, the phenomenon of injury has explained the significant underestimation of coliform bacteria[19-21] as well as false negative results in finished drinking water.[16] These findings clearly indicate a need for more conclusive information in this area.

Other unresolved questions relate to the health effects of bacterial colonization of activated carbon used in the treatment of drinking water. An earlier paper[22] raised this issue by suggesting that activated carbon filters could represent a health hazard. Since that time, a number of reports have identified a variety of bacteria in GAC-treated water, including some that are potentially opportunistic pathogens.[2,5-7] In addition, little attention has been given to the isolation of coliforms in GAC-treated drinking water, although this finding has been reported.[6] However, many bacteria found on GAC have been antagonists that could have suppressed coliform detection,[23-25] and injured coliforms have likely gone undetected because accepted procedures were used that do not enumerate damaged cells.[19-21] Another legitimate health-related question deals with the disinfectant susceptibility of GAC-associated bacteria because of activated carbon's significant chlorine demand[2] and the general observation that turbidity has a deleterious effect on disinfection by chlorine.[26,27] This issue, likewise, is not resolved.

The preceding review of studies dealing with the microbiology of activated carbon in the treatment of drinking water suggests a number of important questions that need to be answered. A useful understanding of how this important technology impacts bacterial water quality awaits this contribution. A listing of these questions follows.

1. What methods are necessary to accurately enumerate GAC-associated bacteria?
2. What is the incidence of GAC breakthrough into finished drinking water?
3. How does the manipulation of operational variables within the treatment plant influence GAC-associated breakthrough?
4. What is the susceptibility of GAC-associated bacteria to disinfection?
5. Can enteric pathogenic bacteria colonize and survive on GAC?
6. How does the growth and physiology of GAC-associated bacteria compare with that of nonassociated cells?

The following sections summarize the findings of a four-year study that addressed these questions. All of the results of this study have been published. For that reason, our findings will simply be summarized and integrated within this discussion with the pertinent references given. More information and detail may be obtained by referring to the individual publications.

RESULTS AND DISCUSSION

Development and Evaluation of Procedures to Desorb Bacteria from GAC

A basic problem underlying many of the previous studies describing the microbiology of GAC in drinking water relates to the need for quantitative enumeration methodology. Failure to desorb and accurately count GAC-associated bacteria seriously underestimates actual numbers and obviously reduces the credibility of conclusions based on such results. For that reason, the first aspect of this project concentrated on the development and evaluation of methods to effectively desorb and enumerate both coliforms and heterotrophic plate count (HPC) bacteria from GAC particles.[28]

Efforts were directed toward the influence of physical factors in the desorption of attached bacteria. These studies revealed that high shear forces were useful, but the resulting heat that was generated reduced bacterial viability. For that reason, homogenization in devices where the suspension could be maintained at 4°C or below by immersion in an ice bath proved most effective. Several suitable instruments are commercially available from manufacturers (such as Brinkmann Instruments, Westbury, NY). In addition, a wide variety of chemical mixtures were tested to determine the formulation that would maximize desorption and prevent bacterial reassociation with GAC surfaces. The mixture that appeared optimal was Tris buffer (0.01 M, pH 7.0), Zwittergent 3–12 ($10^{-6} M$), ethyleneglycol-bis-(beta amino-ethyl ether)-N,N'-tetraacetic acid (EGTA) ($10^{-3} M$) and peptone (0.01%).

Enumeration of HPC bacteria was optimal with R2A medium[18] incubated at 28°C for seven days. For coliform analyses, m-T7 medium[29,30] yielded maximal counts. These procedures were evaluated using samples of GAC taken from filters in operating drinking water treatment plants, as well as GAC inoculated and cultivated in the laboratory.[28] The effectiveness of the desorption technique was evidenced by recovery of 90% of a seeded population. Direct microscopy using the acridine orange epifluorescence method[31] and observations with the scanning electron microscope confirmed the efficiency of bacterial desorption from colonized GAC particles using these techniques.

Bacteria Associated with GAC Particles in Treated Drinking Water

Treated water was sampled from nine operating drinking water systems to test for colonized GAC particles following carbon filtration. This aspect of the study was performed to critically evaluate the question of whether GAC-associated bacteria were penetrating the treatment barriers and inoculating the distribution system. The improved accuracy of bacterial enumeration provided by our desorption procedures was utilized to make the data more definitive than in previous studies.

This task required the development of a mechanism to sample large volumes of treated drinking water for GAC particles. For this purpose, multilayered gauze filters were placed in 45-mm Swinnex filter units connected to sampling taps within treatment plants at points downstream from the filter units. GAC particles were separated from the water by the gauze filters and then treated by our blending and enumeration procedures. A chlorination step was incorporated prior to the blending process to inactivate nonassociated bacteria. Therefore, only GAC-attached bacteria were enumerated in this procedure. In addition, an increase in bacterial CFUs of greater than twofold following the blending process was used as an index of significant GAC colonization. Over 200 samples were sent by refrigerated overnight mail for analysis from the cooperating drinking water treatment facilities. Overall results showed that GAC particles colonized by HPC bacteria were found in 41% of the samples, while 17% had GAC-attached coliforms.[32] These results are summarized in Table 1.

Treatment systems using GAC revealed a twofold greater content of particulates colonized by both HPC bacteria and coliforms than systems using anthracite or sand. The results also indicated that samples collected for the entire filter cycle contained more colonized GAC particles than those collected one hour before or after backwashing. There was no seasonal trend in the occurrence of GAC-associated HPC bacteria, but there were striking seasonal coliform occurrences associated with GAC in the autumn and again during spring months, with 28% of the coliforms exhibiting the fecal biotype. Additional details and information on this aspect of the project are available.[32]

Influence of Operational Variables on Breakthrough of Colonized GAC Particles into Distribution Water

Operational variables were examined to determine if they could be manipulated to minimize bacterial breakthrough or the colonization of GAC filters. Samples were processed from operating drinking water systems, where the variables could be isolated or manipulated. Statistical analysis determined that some of the variables studied correlated with increases in the breakthrough of particle associated bacteria. There was greater breakthrough with higher applied water turbidity (2.9 and 4.3 average NTUs), greater filter depth (samples removed from 2- and 5-foot depths of the filter) and higher filtration rate (2 and 4 gal/min/square foot), although filter age did not influence penetration of colonized particles. In addition, laboratory studies examining the colonization of new filter media by *Klebsiella oxytoca* and field studies revealed that both sand and anthracite were colonized significantly less than GAC, while results with HPC bacteria showed little difference.[33]

Table 1. **Results of Bacterial Analyses of Particles Collected from GAC Treated Drinking Water**

	Heterotrophic Plate Count	Coliform	
		MF	MPN
Total number of samples	198	201	191
Number with GAC-associated bacteria[a]	82 (41.4)[b]	14 (7.0)	33 (17.2)
Meanfold increase	8.6	124.3	24.5
Maximum increase	50.0	1194.0	122.2

[a]Determined by a 2-fold total increase in bacterial numbers in the supernatant following homogenization versus handshaken.
[b]Numbers in parentheses indicate percent of total samples.

Disinfection Susceptibility of Bacteria Attached to GAC

An understanding of the susceptibility of GAC-associated bacteria to disinfection by chlorine is important, since breakthrough of colonized particles occurs in operating drinking water systems. Interest in this consideration also stems from the general observation made by others that particle-associated bacteria tend to be more resistant to disinfectants.[26]

Colonized carbon particles and those with recently adsorbed cells were treated with 2.0 mg/L chlorine, and the microbial population analyzed at timed intervals using the desorption and enumeration methods described earlier. These results were compared with data from identical experiments where nonassociated planktonic bacteria were used.[34] The GAC samples were from operating drinking water treatment plants and laboratory-inoculated material. The unattached bacteria decreased sharply over the first five minutes of chlorine exposure, while the attached cells were virtually unaffected even after one hour, as seen in Figure 1.

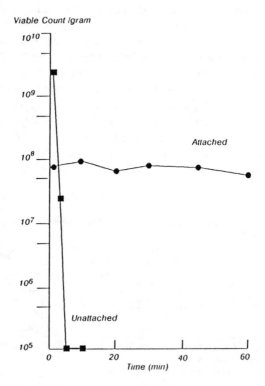

Figure 1. Survival of naturally occurring heterotrophic plate count bacteria exposed to chlorine at 2.0 mg/L for 1 hour (free chlorine residual after 1 hour was 1.7 mg/L).

The free chlorine levels after one hour remained between 1.4 and 1.7 mg/L. Microorganisms examined in these studies included HPC bacteria, coliforms, and waterborne pathogens (*Salmonella typhimurium, Yersinia enterocolitica*, and *Shigella sonnei*), Table 2.

All bacteria examined were highly resistant to disinfection when attached to GAC. Additional experiments indicated that little chlorine interacted with the cells grown on the GAC. Observation of particles with mature biofilms by scanning electron microscopy revealed extensive colonization with deposition of extracellular polymer.[34]

Growth and Persistence of Waterborne Pathogens on GAC

Studies were conducted to determine the colonization, growth, and persistence of enteric pathogenic bacteria on GAC. The health significance of this series of experiments stems from the previous findings that bacteria can penetrate treatment barriers when colonized on GAC particles and are highly resistant to levels of chlorine usually found in drinking water. In addition, routine testing of GAC-treated and chlorinated drinking water effluents from nine operating systems revealed the presence of particle-Associated pathogens and fecal coliforms.[32]

Table 2. Viability and Injury of Enteric Bacteria on GAC with Chlorine Exposure

Species[a]	Time Chlorination (min)	Decrease in log Viability[b]	Percent Injury[c]
Salmonella typhimurium			
Grown on GAC, 2 ppm of chlorine	60	0.37	44
Attached, 2 ppm of chlorine	60	0.50	9.1
Control[d]	5	6	ND[e]
Yersinia enterocolitica			
Grown GAC, 2 ppm of chlorine	60	0.10	16
Attached, 2 ppm of chlorine	60	0.40	16
Control	5	5	ND
Shigella sonnei			
Attached, 2 ppm of chlorine	60	0.14	0
Control	5	5	ND
E. coli			
Grown on GAC, 2 ppm of chlorine	60	0.1	0
Attached, 2 ppm of chlorine	60	0.1	60
Control	5	2.5	ND

[a]Data for cells grown on GAC are the average of two experiments; data for washed cells attached to GAC are the average of three experiments; control cells were washed once and suspended with no carbon.
[b]Log viability calculated by log TLY (nonselective) counts at 0 time minus log TLY counts at 60 min.
[c]Log injury calculated by log TLY-D (selective) counts at 0 time minus log TLY-D counts at 60 min.
[d]Bacteria not attached to GAC.
[e]ND = not determined.

Experiments were carried out using laboratory-scale GAC columns challenged with suspensions of *Yersinia enterocolitica, Salmonella typhimurium,* and enterotoxigenic *E. coli.* GAC samples were removed and examined using the blending procedure described earlier. Bacteria were enumerated using appropriate media and incubation conditions. In the first series of experiments, pathogens were added to columns of sterile carbon receiving sterile river water. The results indicated that the three pathogens could colonize the columns and maintain high population levels during experiments of 14–20 days. These stable pathogen communities were then challenged with unsterile river water, which resulted in the gradual decline of the pathogen populations. In the next studies, mixed populations of the pathogens and native bacteria from a local oligotrophic river were added together to sterile columns of GAC. High levels of both pathogens and nonpathogens became initially established on the GAC. However, pathogen levels gradually declined over time. In the last series of experiments, GAC columns colonized with mature communities of native river bacteria were challenged with pathogens. This approach resulted in the establishment of somewhat lower levels of pathogens that declined more rapidly with time. Typical results are shown in Figure 2. All three of the pathogens used in these experiments reacted similarly.[35]

Growth and Physiological Characteristics of GAC-Associated Coliforms

Information from numerous workers suggests that solid-liquid interfaces represent a microzone of nutrient enrichment in dilute aquatic environments. For that reason and because of the notable adsorptive properties of activated carbon, it was anticipated that GAC would stimulate bacterial growth in drinking water environments. Therefore, this aspect of the project was pursued to examine the comparative physiological properties and growth rates of GAC-associated and planktonic cells.[36]

Experiments were performed to evaluate the comparative growth and physiology of *Klebsiella oxytoca*, isolated from GAC in drinking water, grown attached to GAC and in liquid medium. Laboratory studies showed that when this organism attached to GAC, the growth rate was enhanced more than ten times in the presence of less than 50 mg/L glutamate, a substrate that adsorbed to GAC. No differences were observed in the same concentration range of glucose, which did not adsorb to GAC. Cellular (^{3}H)thymidine uptake was used to estimate DNA biosynthesis. Attached bacteria grown in a minimal nutrient medium containing 20.0 mg/L glutamate took up five times more (^{3}H)thymidine than cells grown in suspension. (^{3}H)Uridine was used as a measure of RNA turnover. Attached cells were shown to assimilate 11 times more (^{3}H)uridine than cells in liquid medium. Cell size measurements were performed by differential filtration. Cells

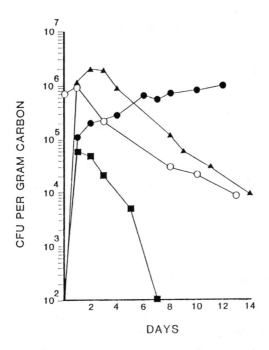

Figure 2. Attachment and persistence of *Y. enterocolitica* 0:8 on GAC columns. Symbols: ● = sterile river water, sterile carbon; ○ = continuation of above, nonsterile river water added; ▲ = nonsterile river water, sterile carbon; ■ = nonsterile river water, precolonized carbon.

grown in a minimal medium with 20.0 mg/L glutamate decreased in size over time, with 62% of the total number passing through a 1.0-μm filter after nine days' incubation. In the same period, 39% of a cell population that was grown on GAC passed through a 1.0-μm filter. These studies indicate that GAC may provide an enhanced interfacial environment for the growth of *K. oxytoca* in the presence of nutrients that are adsorbed upon that surface.

CONCLUSIONS

1. Conventional enumeration procedures underestimate GAC-attached bacterial populations. Homogenization with a mixture of Tris buffer (0.01 M, pH 7), Zwittergent 3–12 ($10^{-6} M$), EGTA ($10^{-3} M$), and peptone (0.01%) at 4°C for 3 min provided optimal desorption. R2A medium incubated at 28°C for seven days and mT7 medium were most efficient in enumerating HPC and coliform bacteria, respectively.

2. HPC and coliform bacteria colonize operational GAC filters, and populated particles are released even from properly maintained filters in drinking water systems.

3. Penetration of populated carbon fines through the treatment barriers was related to increased flow rate, bed depth of GAC filters, and decreased applied water quality. Seasonal influences may also be important.

4. GAC-attached bacteria were not susceptible to chlorine levels commonly used to disinfect potable water.

5. Enteric pathogens can colonize GAC and persist for a varied amount of time depending upon the indigenous HPC population present.

6. GAC-associated coliform bacteria exhibit a greater growth rate and other physiological indices than their planktonic counterparts under controlled laboratory conditions.

ACKNOWLEDGMENTS

We thank Jerrie Beyrodt, Nancy Burns, Diane Matheson, Pamela Blevins, Tom Guza, Sherri Mudri, Anna Moran, Bruce Lapke, and Rich Robisco for technical assistance. E. E. Geldreich and D. J. Reasoner are acknowledged for their helpful suggestions through the course of this project. The cooperation of the participating drinking water treatment facilities is greatly appreciated. This study was supported by funds from the Drinking Water Research Division, U.S. Environmental Protection Agency, Cincinnati, Ohio (grant No. CR 810015).

REFERENCES

1. *Drinking Water and Health, Vol. 1* (Washington, DC: National Academy of Sciences, The National Research Council, 1977).
2. *Drinking Water and Health, Vol. 2* (Washington, DC: National Academy of Sciences, The National Research Council, 1980).
3. United States Environmental Protection Agency. "National Interim Primary Drinking Water Regulations," *Federal Register* 73:59568–59588 (1975).
4. United States Environmental Protection Agency. "National Interim Primary Drinking Water Regulations, Control of Trihalomethanes in Drinking Water, Final Rule," *Federal Register* 44(231) (1979).
5. AWWA Research and Technical Practice Committee on Organic Contaminants. "An Assessment of Microbial Activity on GAC," *J. Am. Water Works Assoc.* 73:447–454 (1981).

6. Brewer, W. S., and W. W. Carmichael. "Microbial Characterization of Granular Activated Carbon Filter Systems," *J. Am. Water Works Assoc.* 71:738–740 (1979).

7. Cairo, P. R., J. McElhaney, and I. H. Suffet. "Pilot Plant Testing of Activated Carbon Adsorption Systems," *J. Am. Water Works Assoc.* 71:660–673 (1979).

8. Schalekamp, M. "The Use of GAC Filtration to Ensure Quality in Drinking Water from Surface Sources," *J. Am. Water Works Assoc.* 71:638–647 (1979).

9. Weber, W. J., Jr., M. Pirbazari, and G. L. Melson. "Biological Growth on Activated Carbon: An Investigation by Scanning Microscopy," *Environ. Sci. Technol.* 12:817–819 (1978).

10. Biton, G., and K. C. Marshall, Eds. *Adsorption of Microorganisms to Surfaces* (New York: John Wiley and Sons, 1980), 439 pp.

11. Marshall, K. C., Ed. *Microbial Adhesion and Aggregation* (Dahlem Konfrenzen) (Berlin: Springer-Verlag, 1984), 432 pp.

12. Robeck, G. G., K. A. Dostal, and R. L. Woodward. "Studies of Modifications in Water Filtration," *J. Am. Water Works Assoc.* 56:198–213 (1964).

13. Syrotynski, S. "Microscopic Water Quality and Filtration Efficiency," *J. Am. Water Works Assoc.* 63:237–245 (1971).

14. Amirtharaja, A., and D. P. Wetstein. "Initial Degradation of Effluent Quality," *J. Am. Water Works Assoc.* 72:518–524 (1980).

15. Zaske, S. K., W. S. Dockins, and G. A. McFeters. "Cell Envelope Damage in *Escherichia coli* Caused by Short Term Stress in Water," *Appl. Environ. Microbiol.* 40:386–390 (1980).

16. McFeters, G. A., M. W. LeChevallier, and J. S. Kippin. "Injured Coliforms in Drinking Water," *Appl. Environ. Microbiol.* 51:1–5 (1986).

17. Ridgway, H. F., and B. H. Olson. "Scanning Electron Microscope Evidence for Bacterial Colonization of a Drinking Water Distribution System," *Appl. Environ. Microbiol.* 41:274–287 (1981).

18. Reasoner, D. J., and E. E. Geldreich. "A New Medium for the Enumeration and Subculture of Bacteria from Potable Water," *Appl. Environ. Microbiol.* 49:1–7 (1985).

19. McFeters, G. A., and A. K. Camper. "Enumeration of Coliform Bacteria Exposed to Chlorine," in *Advances in Applied Microbiology, Vol. 29*, A. I. Laskin, Ed. (New York: Academic Press, Inc., 1983), 177–193.

20. LeChevallier, M. W., and G. A. McFeters. "Causes, Implications and Methods for Enumeration of Injured Coliforms in Drinking Water," *J. Am. Water Works Assoc.* 77:81–87 (1985).

21. LeChevallier, M. W., and G. A. McFeters. "Recent Advances in Coliform Methodology for Water Analysis," *J. Environ. Health* 47:5–9 (1984).

22. Wallis, C., G. H. Stagg, and T. L. Melnick. "The Hazards of Incorporating Charcoal Filters into Domestic Water Systems," *Water Res.* 8:111–113 (1974).

23. Geldreich, E. E., H. D. Nash, D. J. Reasoner, and R. H. Taylor. "The Necessity of Controlling Bacterial Populations in Potable Waters: Community Water Supply," *J. Am. Water Works Assoc.* 64:596–602 (1972).

24. Hutchinson D., R. H. Weaver, and M. Scherago. "The Incidence and Significance of Microorganisms Antagonistic to *Escherichia coli* in Water," *J. Bacteriol.* 45:29 (1943).

25. LeChevallier, M. W., and G. A. McFeters. "Interactions Between Heterotrophic Plate Count Bacteria and Coliform Organisms," *Appl. Environ. Microbiol.* 49:1338–1341 (1985).
26. LeChevallier, M. W., T. M. Evans, and R. J. Seidler. "Effect of Turbidity on Chlorination Efficiency and Bacterial Persistence in Drinking Water," *Appl. Environ. Microbiol.* 42:159–167 (1981).
27. Tracy, H. W., V. M. Camarena, and F. W. Wing. "Coliform Persistence in Highly Chlorinated Waters," *J. Am. Water Works Assoc.* 58:1151–1159 (1966).
28. Camper, A. K., M. W. LeChevallier, S. C. Broadaway, and G. A. McFeters. "Evaluation of Procedures to Desorb Bacteria from Granular Activated Carbon," *J. Microbiol. Methods* 3:187–198 (1985).
29. LeChevallier, M. W., S. C. Cameron, and G. A. McFeters. "New Medium for the Improved Recovery of Coliform Bacteria from Drinking Water," *Appl. Environ. Microbiol.* 45:484–492 (1982).
30. LeChevallier, M. W., P. E. Jakanoski, A. K. Camper, and G. A. McFeters. "Evaluation of m-T7 Agar as a Fecal Coliform Medium," *Appl. Environ. Microbiol.* 48:371–375 (1984).
31. Daley, R. J. "Direct Epifluorescence Enumeration of Native Aquatic Bacteria: Uses, Limitations and Comparative Accuracy," in *Native Aquatic Bacteria: Enumeration, Activity and Ecology,* J. W. Costeron and R. R. Colwell, Eds. (Philadelphia, PA: American Society for Testing and Materials, 1977), 29.
32. Camper, A. K., M. W. LeChevallier, S. C. Broadaway, and G. A. McFeters. "Bacteria Associated With Carbon Particles in Drinking Water," *Appl. Environ. Microbiol.* 52:434–438 (1986).
33. Camper, A. K., S. C. Broadaway, M. W. LeChevallier, and G. A. McFeters. "Operational Variables and the Release of Colonized Granular Activated Carbon Particles in Drinking Water," *J. Am. Water Works Assoc.* 79:70–74 (1987).
34. LeChevallier, M. W., T. H. Hassenauer, A. K. Camper, and G. A. McFeters. "Disinfection of Bacteria Attached to Granular Activated Carbon," *Appl. Environ. Microbiol.* 48:918–923 (1984).
35. Camper, A. K., M. W. LeChevallier, S. C. Broadaway, and G. A. McFeters. "Growth and Persistence of Pathogens on Granular Activated Carbon Filters," *Appl. Environ. Microbiol.* 50:1378–1382 (1985).
36. Davies, D. G., and G. A. McFeters. "Growth and Comparative Physiology of *Klebsiella oxytoca* attached to GAC Particles and in Liquid Media," *Microbial Ecology* (in press, 1987).

SECTION VII

Recent Developments

CHAPTER 20

Biodegradation Processes to Make Drinking Water Biologically Stable

Bruce E. Rittmann, Environmental Engineering and Science, University of Illinois at Urbana-Champaign, Urbana, Illinois

INTRODUCTION TO BIOLOGICAL INSTABILITY

Recent findings from Europe and the United States show that the presence of biodegradable materials in a drinking water can bring about significant problems during treatment and distribution of the water. The biodegradable materials most normally found in drinking water supplies are ammonia nitrogen[1] and organic compounds,[2] the majority of which are naturally occurring humic materials. A water containing nonnegligible concentrations of biodegradable materials is termed *biologically instable*.[1]

The problems directly occurring during treatment of a biologically instable water are excessive backwashing caused by bacterial growth in rapid filters and the need for heavy chlorination to prevent biological slime formation in filters and other processes. In the distribution system, biodegradation of the instability materials can create tastes and odors, increases in turbidity and plate counts, increased rates of corrosion, and the need for still more chlorination to try to mitigate the other problems.[1] Chlorination in excess of that needed for pathogen disinfection creates its own set of problems: formation of trihalomethanes (THMs) and other chlorinated by-products,[3,4] increased corrosion,[5] and medicinal tastes and odors.[6] Of course, extra chlorination is an economic cost by itself. Therefore, the presence of biodegradable materials in a drinking water and the common means to forestall its effects—i.e., chlorination—have water quality, aesthetic, operational, and economic costs.

MAKING THE WATER BIOLOGICALLY STABLE

A drinking water from which nearly all of the biodegradable material has been removed is called *biologically stable*.[1] The most logical and economical approach to making a water biologically stable is to include a biological treatment process as part of the overall treatment scheme. Placing biological treatment as one of the first processes is most advantageous, because early removal of biodegradable materials guards against slime buildup in conventional processes, such as coagulation and rapid filtration, and eliminates the need for chlorination beyond that required for pathogen destruction.

Today, biological treatment is relatively common in several European countries. The Federal Republic of Germany and The Netherlands commonly use "natural" biological treatment as part of bank or dune filtration, a form of pretreatment before more conventional processes.[7,8] In France and to a lesser extent in Great Britain, engineered biological reactors are employed very early in drinking water treatment.[1,9] Again, conventional processes follow for removal of turbidity, hardness, pathogens, and synthetic organic chemicals.

In many European countries, as well as in the United States, ozonation prior to granular activated carbon (GAC) filtration enhances biodegradation in the GAC beds. Sometimes this process coupling is called *biological activated carbon* (BAC).[10-12] Having biodegradation in GAC filters serves to make the water more biologically stable and can extend the life of the activated carbon. In addition, rapid sand filters can effect removals of instability compounds when chlorination is not practiced ahead of the filters,[13-15] and slow sand filters are well-known for their biological activity.[13,16] Rittmann[2] and Rittmann and Snoeyink[1] reviewed the evidence for biological removal of organic and inorganic instability materials in the range of drinking water processes.

Biological Treatment Processes

All biological processes used to make biologically stable drinking water share certain characteristics. Those characteristics are low concentrations of biodegradable material, mixed substrates, predominance of attached biomass, and aerobic conditions.

Low Concentrations

The concentrations of biodegradable compounds in drinking water supplies always are low in comparison to the concentrations found in wastewaters or in biological systems used in industrial fermentations. The main inorganic source of biological instability is ammonia nitrogen, and it seldom is found at more than 3 mg N/L. However, concentrations as small as 0.25

mg N/L present significant problems of instability.[1] Fortunately, biological treatment is capable of reducing NH_4^+–N concentration to 0.01 mg N/L or lower, as long as surface loadings are not too high and dissolved oxygen is available.[1,9,17,18]

The other major source of instability is biodegradable organic matter. Although the dissolved organic carbon (DOC) of most water supplies varies from around 1 mg C/L to around 10 mg C/L, the easily degradable portion usually is only a small fraction of the DOC. For example, van der Kooij et al.[19–22] developed a special bioassay procedure that uses specialized bacterial strains to measure the very low concentrations of biodegradable organic material. The result of the assay is called the AOC, or assimilable organic carbon, because its degradation allows bacterial growth. Raw waters in The Netherlands had 7 to 500 μg AOC/L, but that represented only 0.1 to 8.5% of the DOC. Ozonation increased the AOC by 10 to 20 times.

Work in The Netherlands[20] indicates that the easily assimilable organic carbon (AOC) need be only about 25 μg C/L to cause growth problems in the distribution system. Thus, the relevant concentration of biodegradable organic material is much less than 1 mg C/L when achieving a biologically stable drinking water is the goal.

A very recent report by Servais et al.[23] described a different technique for measuring biodegradable DOC. Using a small natural inoculum to assay degradable DOC, Servais et al.[23] found that 11 to 39% of the DOC in raw water was degradable and that the concentrations were 2.0 to 4.9 mg/L.

Mixed Substrates

Besides being present in low concentrations, the biodegradable substrates must also be characterized as being made up of many different components. Within the organic fraction, the few mg C/L of DOC can comprise hundreds of different compounds, each present at concentrations from as high as the tens of μg C/L to as low as the ng C/L level. Such a milieu is called a multiple-substrate system, and it appears that many (and probably most) of the multiple substrates are utilized simultaneously by a few species of heterotrophic microorganisms that can survive and function under the stringent conditions of drinking water treatment. Simultaneous utilization of multiple and varied substrates is a strategy employed by a class of bacteria, known as oligotrophs, to sustain themselves when substrate concentrations are very low.[24,25]

Recent research by Namkung and Rittmann[26,27] has shown that simultaneous use of multiple substrates offers a kinetic advantage to microorganisms, especially those that are present in biofilm processes (see below). Simultaneous utilization of numerous substrates creates an "enhancement" effect which allows greater biomass accumulation and faster individual organic removal, in comparison with the same concentration of a single substrate. Although the current results[26,27] need significant theoretical and modeling advancements, they imply that a very low concentration of organic instability can be achieved by biodegradation for the multiple-substrate systems actually present in drinking water.

The organic components of instability also are primarily natural polymers, or humic substances. Generally thought to be the relatively refractory remains from the decay of natural biomass production (e.g., trees, leaves, and algae), humic materials are moderately large polymers containing a mixture of substituents, such as carboxylic acids, amines, phenolics, and carbohydrates. Their large molecular weights and complex structures give humic materials relatively slow biodegradation kinetics.[2] Nonetheless, humic materials are biodegradable.[2] For instance, Namkung and Rittmann[28] used a peat fulvic acid as the sole carbon and electron source to grow biofilms in lab-scale reactors similar to drinking water systems. The degradation of the peat fulvic acid (about a 10% decrease in 1 mg/L of DOC) allowed accumulation of a visible biofilm that was capable of degrading phenol, naphthalene, geosmin, and MIB (methylisoborneol).

Having ammonium nitrogen and organic substrates present together provides a second level of substrate multiplicity. That is, two distinctly different groups of microorganisms are required to oxidize the organic materials and the ammonium nitrogen. The bacteria that oxidize organic compounds also utilize the carbon in the organic substrates as their carbon source. On the other hand, the nitrifying bacteria, which oxidize the ammonium nitrogen to nitrate nitrogen, are autotrophs, because they must reduce and incorporate inorganic carbon (e.g., bicarbonate) for a carbon source. The use of an inorganic carbon source is quite expensive for the autotrophs and means that their maximum growth rates are slower than for the heterotrophs.[18] Thus, removal of ammonium and organic instability components sets up a form of competition between heterotrophs and autotrophs. Proper process design requires that neither group be put at a competitive disadvantage, but that both groups can accumulate and function together.[18] Current research in the author's laboratory is addressing the competitive and synergistic interactions between the heterotrophs and the nitrifying autotrophs.

Attached Growth

The third aspect of biological treatment to prepare a biologically stable water is that all the processes are of the biofilm type. In other words, the bacteria are attached as a naturally occurring film on solid media. In bank and dune filtration, the attachment medium is the sandy grains of the bank or dune. In the engineered systems, the bacteria are attached to small rocks, stones of pozzolana, particles of expanded clay, fluidized particles of sand or silt, or plastic media. In all cases, the water flows by quickly. For example, liquid detention times in the engineered reactors are only a few minutes.[1,9] However, through attachment, the microorganisms can have retention times measured in weeks.[18]

Biofilm processes are particularly advantageous for the drinking water application. First, biofilm growth is advantageous to the microorganism, because it can maintain a long retention time in the reactor, where it is continually exposed to a fresh supply of substrate (albeit at a low concentration). Thus, oligotrophic microorganisms tend to form biofilms naturally. Second, attachment offers a simple and reliable means to retain the organisms within the reactor; problematic separations operations such as sedimentation are avoided. Third, the combination of high cell retention with low liquid detention times makes it possible to have relatively small process volumes. In many cases, retrofit of a biological process into a conventional facility is feasible.

Because biofilms are attached to surfaces, the design criteria for a biofilm process are in terms of surface characteristics. A high specific surface area is desired to maximize the volumetric reaction rates; specific surface areas in excess of 200 m^{-1} are typical. The substrate loading is given as a surface loading, or a flux. Typical drinking water fluxes are 0.4 kg NH_4^+–$N/10^3$ m^2/day for nitrification. Rittmann[19] recently summarized loading criteria for biofilm processes.

Aerobic Conditions

Finally, the biological treatment processes are aerobic. Aerobic conditions are needed to allow nitrification and to prevent problems of taste, odors, and metals solubilization associated with anaerobic conditions. When the total oxygen demand exceeds a few mg/L, preoxygenation or aeration in the reactor is needed. One convenient means of preoxygenation is ozonation, which also makes the humic materials more biodegradable.

CONCLUSION

Biological treatment can and should play an important role in the production of a water that is free from biohazards. In this chapter, the aspect of achieving biological stability was stressed. When a water is biologically stable, it is free from organic and inorganic constituents which can biodegrade in the distribution system. This stability means that excessive chlorination need not be practiced, thereby abating the formation of THMs, TOX, tastes, and odors which are associated with chlorination.

REFERENCES

1. Rittmann, B. E., and V. L. Snoeyink. "Achieving Biologically Stable Drinking Water," *J. Am. Water Works Assoc.* 76:106–114 (1984).
2. Rittmann, B. E. "Biological Processes and Organic Micropollutants in Treatment Processes," *Sci. Total Environ.* 47:99–113 (1985).
3. Rook, J. J. "Chlorination Reactions of Fulvic Acids in Natural Waters," *Environ. Sci. Technol.* 11:478 (1977).
4. Johnson, J. D., and J. N. Jensen. "THM and TOX Formation: Routes, Rates, and Precursers," *J. Am. Water Works Assoc.* 78(4):156–162 (1986).
5. Reiber, S. H., J. F. Ferguson, and M. M. Benjamin. "Corrosion Monitoring and Control in the Pacific Northwest," *J. Am. Water Works Assoc.* 79(2):71–74 (1987).
6. Burttschell, R. H., A. A. Rosen, F. M. Middleton, and M. B. Ettinger. "Chlorine Derivatives of Phenol Causing Taste and Odor," *J. Am. Water Works Assoc.* 51:205 (1959).
7. Piet, G. J., and B. C. J. Zoeteman. "Organic Water Quality Changes During Sand Bank and Dune Filtration of Surface Waters in The Netherlands," in *Water Reuse of Groundwater Recharge* (Sacramento, CA: California State Water Resources Control Board, 1980), 195–215.
8. Sontheimer, H. "Experience with River Bank Filtration Along the Rhine River," in *Wastewater Reuse for Groundwater Recharge* (Sacramento, CA: California State Water Resources Control Board, 1980), 195–214.
9. Snoeyink, V. L., and B. E. Rittmann. "Drinking Water and Wastewater Treatment to Remove Ammonia," in *Innovations in the Water and Wastewater Fields,* E. A. Glysson, D. E. Swan, and E. J. Way, Eds. (Stoneham, MA: Butterworth, 1985), 401-410.
10. Rice, R. G., L. Gomella, and G. W. Miller. "Rouen, France Water Treatment Plant," *Civil Eng.* (May 1978):76–82.
11. Jekel, M. "Biological Treatment of Surface Waters in Activated Carbon Filters," Engler-Bunte Institute, University of Karlsruhe, Karlsruhe, Federal Republic of Germany, 1977.
12. Gomella, C., and O. Versanne. "Nitrification biologique et affinage d'une eau de forage," *Tech. et Sci. Municipules* (1980):211.
13. Eberhardt, M., S. Madsen, and H. Sontheimer. "Investigations of the Use of Biologically Effective Activated Carbon Filters in the Processing of Drinking Water," U.S. Environmental Protection Agency, EPA-TR-77–503, Washington, DC, 1977.
14. Sontheimer, H. "The Mulheim Process," *J. Am. Water Works Assoc.* 70:393 (1978).
15. Glaze, W. H., J. L. Wallace, K. L. Dickson, D. P. Wilcox, K. R. Johansson, E. Chang, A. W. Busch, B. G. Scalf, R. K. Noack, and O. P. Smith, Jr. "Evaluation of Biological Activated Carbon for Removal of Trihalomethane Processors," Municipal Environmental Research Lab, U.S. Environmental Protection Agency, Cincinnati, OH, 1982.

16. "Activated Carbon in Drinking Water Technology," in *Report of the Study Group on Activated Carbon,* J. C. Kruithof and R. C. van der Leer, Eds. (Nievweigen, The Netherlands: KIWA, 1983).
17. Rittmann, B. E. "Potential for Controlling SOCs Through Biotransformations," in *AWWA Seminar Proceedings. Treatment Processes for Control of Synthetic Organic Chemicals* (Denver, CO: American Water Works Association, 1987), 233-246.
18. Rittmann, B. E. "Aerobic Biological Treatment," *Environ. Sci. Technol.* 21:128–136 (1987).
19. Van der Kooij, O. A. Visser, and W. A. M. Hijnen. "Determining the Concentration of Easily Assimilable Organic Carbon in Drinking Water," *J. Am. Water Works Assoc.* 74:540 (1982).
20. Van der Kooij, D. Personal communication, June 1985.
21. Van der Kooij, D., J. P. Orange, and W. A. M. Hijnen. "Growth of *Pseudomonas aeruginosa* on Tap Water in Relation to Utilization of Substrate at Concentrations of a few Micrograms per Liter," *Appl. Environ. Microbiol.* 44:1086-1095 (1982).
22. Van der Kooij, D., and W. A. M. Hijnen. "Substrate Utilization by an Oxalate-Consuming *Spirillum* Species in Relation to Its Growth in Ozonated Water," *Appl. Environ. Microbiol.* 47:551–559 (1984).
23. Servais, P., G. Billan, and M.-C. Hascoet. "Determination of the Biodegradable Fraction of Dissolved Organic Matter in Waters," *Water Res.* 21:445–450 (1987).
24. Hirsch, P., M. Bernhard, S. Cohen, J. Ensign, H. Jannasch, A. Koch, K. Marshall, A. Martin, J. Poindexter, S. Rittenberg, O. Smith, and H. Veldkamp. "Life Under Conditions of Low Nutrient Concentration, Group Report," in *Strategies for Microbial Life in Extreme Environments,* M. Shilo, Ed. (New York: Dahlem Konferenzen, Verlag Chemie, 1979), 357–372.
25. Poindexter, J. S. "Oligotrophy. Feast and Famine Existence," *Adv. Microbiol. Ecol.* 5:63–89 (1981).
26. Namkung, E., and B. E. Rittmann. "Modeling Bisubstrate Removal by Biofilms," *Biotechnol. Bioeng.* 24:269–278.
27. Namkung, E., and B. E. Rittmann. "Evaluation of Bisubstrate Secondary Utilization Kinetics by Biofilms," *Biotechnol. Bioeng.* 24:335–342 (1987).
28. Namkung, E., and B. E. Rittmann. "Removal of Taste and Odor Compounds by Humic-Substances-Grown Biofilms," *J. Am. Water Works Assoc.* 79(7):107 (1987).

CHAPTER 21

Application of Gene Probes to the Detection of Enteroviruses in Groundwater

Aaron B. Margolin, Kenneth J. Richardson, Ricardo DeLeon, and Charles P. Gerba, Departments of Microbiology/Immunology and of Nutrition and Food Sciences, University of Arizona, Tucson, Arizona

INTRODUCTION

Rapid and low-cost methods for the detection of enteric viruses in water have been sought by water virologists for many years. Virus detection in water requires passage of 400–1000 liters of water being sampled through a filter to which the viruses adsorb. Viruses eluted from these filters are then assayed in animal cell culture.[1] Although cell culture techniques are now relatively simple to perform, this assay system does have several drawbacks. One drawback is that incubation periods of three days to six weeks may be required before cytopathogenic effects (CPE) are observed. A second problem with this assay system is the lack of one cell line that will permit the replication of all enteric viruses.[1] In addition, some enteric viruses, such as hepatitis A virus, do not exhibit cytopathogenic effects in cell culture.[2] Other viruses, such as Norwalk virus, have not yet been grown in cell culture. In addition, cell culture is very expensive and can cost between $300 to $750 per sample.

More rapid techniques that are commonly used in the human clinical laboratory, such as fluorescent antibody (FA) or radioimmunoassay (RIA), lack the needed sensitivity necessary to detect the low numbers of viruses found in contaminated water.[3-5] Since low numbers of viruses in water are believed significant in the spread of waterborne disease, methods must be capable of detecting as few as 1–10 infectious viruses in 100–1000 liters of drinking water.[6]

Current advances in DNA technology using gene probes now provide a method for identifying the genes of any organism. Gene probes are small

265

pieces of complementary DNA or RNA that have been labeled with either an isotope or a nonradioactive compound such as biotin.[7] The development of gene probes against enteric viruses now makes possible the rapid detection of enteric viruses in water. Up to 96 concentrated water samples can be probed at one time providing results within 72 hours.

MATERIALS AND METHODS

Cells and Viruses

Poliovirus type 1 (LSc) was assayed and grown in the Buffalo Green Monkey (BGM) continuous cell line. All assays were by the plaque-forming unit (PFU) methods.[2] Hepatitis A virus (enterovirus type 72) strains HAS 15, CR326, and HM 175 were grown and assayed in the FRhK6 continuous cell line. The hepatitis A virus (HAV) was quantitated by radioimmunofocus assay.[8]

Probes and Hybridizations

Two different cDNA probes were used for enterovirus detection. A poliovirus type 1 (Mahoney) cDNA probe (base pairs 115–7440) inserted into the Pst 1 site of the plasmid pBR322 was provided by Drs. Rancaniello and Baltimore. The second probe contained the first 1380 base pairs from the 3′ end of the hepatitis A virus. The poliovirus and hepatitis A virus cDNA probes were grown in transformed *Escherichia coli* HB-101. The cDNA probes were extracted from the *E. coli* and isolated on a cesium chloride/ethidium bromide gradient as previously described in Maniatas et al.[7]

The probes were labeled with ^{32}P dCTP and ^{32}P dATP (specific activity 3000 Ci/mmole) using nick translation to a specific activity of 2.0×10^9 cpm/μg of DNA or greater. The entire pBR322 plasmid along with the viral cDNA insert was used as the probe rather than the insert alone.

Prehybridizations and hybridizations were done at 44°C in sealable plastic bags according to the procedures of Thomas.[9] Approximately 1.0×10^7 counts were added to each hybridization bag and hybridizations were carried out for 24–36 hours in a water bath with constant agitation. Hybridization membranes were washed in a $2\times$ SSC ($0.3 M$ sodium chloride, $0.03 M$ sodium citrate) solution at room temperature for 10 minutes. This was then followed by a second wash using $2\times$ SSC, 1% SDS (sodium dodecyl sulfate) at 52°C for 30 minutes. A final wash of $0.2\times$ SSC was done at room temperature for 15–30 minutes. Results were visualized by autoradiography for 24–36 hours at −70°C.

Sample Analysis

Samples were first centrifuged to clarify and remove any large particles. To liberate the viral genomes, samples were originally phenol/chloroform extracted according to the methods of Maniatas et al.[7] This was followed by two chloroform extractions and then by a water-saturated ether extraction. Dissolved ether was removed by bubbling air through the mixture until traces of ether were gone.

Further research done in our laboratory demonstrated that the genome of poliovirus could either be liberated or exposed from the viral protein coat by heating the sample to 65°C for 30 minutes. RNasin was added prior to heating the sample to inhibit RNase activity and prevent RNA degradation from occurring once the genome had been liberated and/or exposed. Water samples (where indicated) were processed in this manner to liberate the viral genome rather than using a phenol/chloroform extraction.

RESULTS

Table 1 demonstrates that poliovirus and hepatitis A virus were detected in seeded tap water with sensitivities equal to the PFU and RIFA assays. To determine if beef extract used in viral elution would interfere with hybridization, 3% beef extract seeded with poliovirus and beef extract without poliovirus was concentrated by organic flocculation and then assayed by the PFU method and the gene probe assay for the presence of virus. Table 2 indicates that beef extract does not interfere with the sensitivity of the gene probe assay nor does it produce false positive results. Poliovirus was detected with equal sensitivities in both assay systems. Table 3 indicates that the sensitivity of poliovirus detection in seeded tap water was equal for the PFU method and the phenol/chloroform extractions, but there was an increased sensitivity in virus detection for the heat-treated sample. Table 4 shows the results of the different groundwater samples that were assayed for virus. Phenol/chloroform extraction of samples or heat treatment of samples are indicated in the table.

DISCUSSION

The results of this research describe the development of a cDNA probe capable of detecting as few as 1 PFU of poliovirus or 1 RIFA unit of hepatitis A virus within 72 hours. Beef extract, which was used to elute viruses from filters did not seem to interfere with the sensitivity of the assay nor did it create false positive results. Phenol/chloroform extractions liberated viral genomes and permitted the detection of viral RNA with sensitivities approxi-

Table 1. Sensitivity of the Dot Blot Assay for Hepatitis A Virus and Poliovirus Detection

Virus	Titer	Dot Formation (Virus Dilutions)					
		10^{-3}	10^{-4}	10^{-5}	10^{-6}	10^{-7}	10^{-8}
Hepatitis A							
HAS 15	9.5×10^5 RIFA/mL	+	+	+	+	−	−
CR 326	2.0×10^6 RIFA/mL	+	+	+	+	−	−
HM 175	1.3×10^6 RIFA/mL	+	+	+	+	−	−
Polio 1							
LSc	1.2×10^4 RIFA/mL	+	+	−	−	ND[a]	ND

[a]ND = not done.

Table 2. Effects of Beef Extract on the Dot Blot Assay for Poliovirus Detection

Sample	Poliovirus PFU/mL	Dot Formation (dilutions)			
		10^{-2}	10^{-3}	10^{-4}	10^{-5}
Beef Extract Concentrate	4.2×10^4	+	+	+	−
Beef Extract	0	−	−	−	−

Table 3. Comparison of Phenol/Chloroform Extraction and Heat Treatment for the Detection of Poliovirus

Method	Dilution Series of Poliovirus						
	10^0	10^{-1}	10^{-2}	10^{-3}	10^{-4}	10^{-5}	10^{-6}
PFU	+	+	+	+	+	−	−
Phenol/chloroform extraction	+	+	+	+	±	−	−
Heat treatment	+	+	+	+	+	±	−

mately equal to tissue culture. Heat treatment of the samples seemed to increase the sensitivity of the assay, making it more sensitive than tissue culture.

The gene probe assay does not first require virus growth in cell culture. This allows samples to be probed directly and decreases the assay time. It also allows for the detection of viruses that may not be infectious, but still retains their genome. However, for untreated groundwater used for drinking water, this test is ideal, because it can screen many samples in a short time to determine if virus is present. Untreated groundwater should not contain any viruses, and hence the gene probe assay can be used as a rapid assay to determine if a sample is contaminated with viruses. Samples which were positive by gene probe assay but negative by tissue culture may have contained viruses which were below the detection limit of tissue culture but were

Table 4. Detection of Naturally Occurring Enteric Viruses in Water Using Gene Probes vs Tissue Culture

Sample	Tissue Culture (CPE)	Gene Probe Phenol/Chloroform	Gene Probe Heat Treatment
1	+	+	+
2	+	+	+
3	+	+	+
4	−	+	+
5	−	+	+
6	−	+	+
7	−	−	−
8	−	−	−
9	−	ND[a]	+
10	+	ND	+
11	+	ND	+
12	+	ND	+
13	+	ND	−

[a]ND = not done.

detected by the gene probe assay. Also, the samples may have contained inactive viruses which could not be replicated in cell culture but were detected by the gene probe assay. Since the samples in Table 4 are undisinfected drinking water samples, these samples should not contain any viruses. Sample 13 (Table 4) indicates tissue culture assay was positive, while gene probe assay was negative. Research in our laboratory has shown that the poliovirus cDNA probe will cross-hybridize with other enteric viruses such as coxsackie B and echoviruses but does so with almost a two-log reduction in sensitivity when compared to tissue culture. Such viruses may be detected by tissue culture but go undetected by the gene probes used in this study.

Upon comparison with tissue culture, the gene probe assay is more rapid and sensitive than cell culture. Phenol/chloroform extractions were one original drawback of the gene probe assay. These extractions required the user to be exposed to potentially harmful organic solvents and were often long and tedious when very proteinatious samples were encountered. Heat treatment of the sample increases the simplicity of the gene probe assay. Samples are treated by adding protease K and then incubating at 65°C for 30 minutes. Prior to this, RNasin is added to help prevent RNA degradation.

Current costs of water analysis for virus contamination can exceed $500 per sample and can take as long as 3–6 weeks for results. The gene probe assay was reliable and sensitive and provided results within 72 hours. Also, the gene probe assay will reduce the cost of testing water to under $150. This type of low-cost sensitive assay will permit water utility companies to monitor for the presence of viruses.

REFERENCES

1. Bitton, G. *Introduction to Environmental Virology* (New York: John Wiley and Sons, 1980).
2. Sobsey, M. D. "Methods for Detecting Enteric Viruses in Water and Wastewater," in *Viruses in Water,* Berg, Bodily, Lennette, Melnick, and Metcalf, Eds. (Washington: American Public Health Association, 1976).
3. Halonen, P., and O. Meurman. "Radioimmunoassay in Diagnostic Virology," in *New Developments in Practical Virology* (New York: Alan Liss, Inc., 1982).
4. Harris, C. C., R. H. Yolken, and H. Krokam. "Ultrasensitive Enzymatic Radioimmunoassay; Application to Detection of Cholera Toxin and Rotavirus," *Proc. Natl. Acad. Sci. USA* 10:5336–5339 (1979).
5. Herrman, J. E., R. M. Hendry, and M. F. Collins. "Factors Involved in Enzyme-Linked Immunoassay of Viruses and Evaluation of the Method for Identification of Enteroviruses," *J. Clin. Micro.* 10:210–217 (1979).
6. Sproul, O. J. "Public Health, Financial and Practical Considerations of Virological Monitoring and Quality Limits," *Water Sci. Technol.* 15:33–41 (1983).
7. Maniatas, T., E. F. Fritsch, and J. Sanbrook. *Molecular Cloning* (New York: Cold Springs Harbor Laboratory, 1982).
8. Lemon, S. M., L. N. Binn, and R. H. Marchwicki. "Radioimmuno Focus Assay for Quantitation of Hepatitis A Virus in Cell Culture," *J. Clin. Microbiol.* 17:834–839 (1983).
9. Thomas, P. S. "Hybridization of Denatured RNA and Small DNA Fragments Transferred to Nitrocellulose," *Proc. Natl. Acad. Sci. USA* 77:5201–5208 (1980).

CHAPTER 22

New Halamine Water Disinfectants

S. D. Worley, D. E. Williams, S. B. Barnela, and L. J. Swango,
Departments of Chemistry and Microbiology, Auburn University, Auburn, Alabama

INTRODUCTION

There is a definite need for a new water disinfectant which exhibits long-term stability and does not react appreciably with organic impurities in water to form toxic trihalomethanes. Work has been progressing in these laboratories aimed toward the development of such a disinfectant. Thus far, two classes of compounds, N-halooxazolidinones and N,N'-dihaloimidazolidinones, which are organic chloramines, have been prepared and tested here for stability and disinfection efficacy under a variety of conditions of pH, temperature, and water quality. The structures of these compounds are shown in Figure 1.

Compound I (3-chloro-4,4-dimethyl-2-oxazolidinone) was first prepared and shown to have bactericidal properties by Bodor et al.[1,2]; this compound has since been studied thoroughly in these laboratories.[3-14] It will be further addressed in this work in comparison with the other compounds shown in Figure 1. Compound IB (3-bromo-4,4-dimethyl-2-oxazolidinone) was also first prepared by Bodor et al., but it was evaluated only as a brominating agent.[15] The three 1,3-dihalo-4,4,5,5-tetramethyl-2-imidazolidinone compounds (A, AB, and ABC in Figure 1) have been synthesized for the first time in these laboratories.[16] This chapter will compare the five N-halamine compounds as stable disinfectants for water applications.

EXPERIMENTAL

The syntheses, purification procedures, and structure identification of the five N-halamine compounds shown in Figure 1 have been outlined previ-

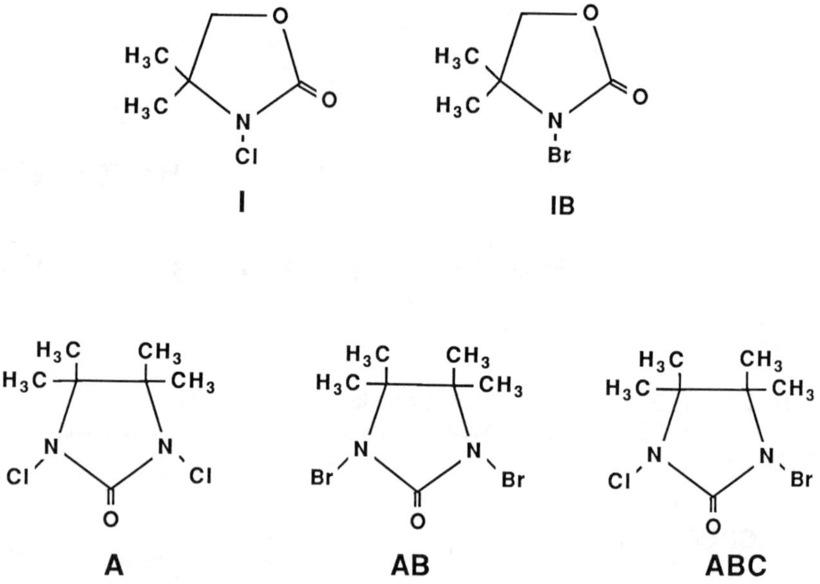

Figure 1.   Structures of the disinfectant compounds in this study.

ously.[1,2,16] In this work two types of testing for the compounds were per-formed—evaluations of the stabilities, and disinfection efficacies against a variety of organisms, in water as a function of pH (4,5, 7.0, 9.5), tempera-ture (2, 22, 37°C), and water quality (demand-free [DFW] and synthetic-demand [SDW]). The synthetic-demand water contained inorganic salts, bentonite clay, humic acid, horse serum, and dead yeast cells and was buff-ered at pH 9.5 and held at 4°C as outlined previously.[7,12] The pH, tempera-ture, and water quality conditions employed were specified by the sponsors of this research.

In the stability experiments each compound was dissolved in water (either DFW or SDW) to a concentration of 10 mg/L total chlorine or its molar equivalent in total bromine or total oxidant. For the DFW experiments, the solutions were stored in flasks vented to the atmosphere through filters con-taining activated charcoal and a desiccant. These solutions were sampled weekly over a period of six weeks, with the total halogen concentration determined by iodometric titration. For the SDW experiments, the solutions were sampled frequently over a period of 60–100 hr. All solutions were maintained at a prescribed temperature before sampling by use of a tempera-ture-controlled water bath.

For the disinfection efficacy experiments, solutions (DFW or SDW) con-taining the organisms at a final concentration of 10^6 cfu/mL were used to challenge each disinfectant compound. Aliquots were withdrawn frequently over a period of several hours, quenched by sodium thiosulfate, and assessed for survivorship. In one experiment *Staphylococcus aureus* was used to repetitively challenge a 2.5 mg/L total chlorine (or its molar equivalent in total bromine or total oxidant) solution of the various disinfectants over a period of four months. In this experiment disinfectant was added only at the beginning. In these studies the organisms employed were: *Staphylococcus aureus* ATCC 25923, *Shigella boydii* ATCC E9207, *Entamoeba invadens* IP-1 strain (trophozoites), *Giardia lamblia* Portland strain (trophozoites), *Giardia lamblia* cysts, *Legionella pneumophila* ATCC 33152, and rotavirus SA-11 strain.

RESULTS AND DISCUSSION

The stabilities of the several disinfectant compounds in water as a function of pH, temperature, and water quality are shown in Figures 2–6. The stability of free chlorine ($Ca(OCl)_2$, H) under these conditions is shown also for comparison. Figure 2 illustrates that at pH 7.0, 22°C, demand-free water, that all of the N-halamine compounds were considerably more stable than free chlorine. The stabilities of I and A under these conditions were the greatest (the apparent difference between I and A is not significant statistically). The N-bromamines were less stable than their N-chloramine analogs, with the mixed bromochloroimidazolidinone (ABC) being intermediate in stability. Figure 3 shows that an increase of temperature to 37°C at pH 7.0 in demand-free water caused the largest effect (loss of stability) for the N-bromamines and that compound A was affected the least. In Figure 4 it is evident that an increase in pH to 9.5 at 22°C in demand-free water affected the stability of the oxazolidinone ring (I and IB) much more than that for the imidazolidinone ring (A, AB, and ABC). This was dramatized by an increase in temperature to 37°C at pH 9.5 in demand-free water (Figure 5), as compound I became very unstable, and only compound A was significantly more stable than free chlorine under these conditions. Figure 6 shows that in synthetic-demand water in the presence of heavy organic load, A was clearly the most stable disinfectant compound. The compounds containing bromine were decomposed rapidly in the SDW solution, as was free chlorine. The apparent leveling of the free chlorine stability curve after several hours represents the formation of chloramines from the reaction of free chlorine with nitrogenous material in the organic load; these chloramines are not biocidal. Compound ABC lost 50% of its total halogen rapidly and then became stable. The bromine moiety was lost, leaving the monochloroimidazolidinone, which is biocidal, as will be demonstrated later.

The bactericidal efficacies of the N-halamine compounds against *S. aureus* and *Sh. boydii* are shown in Figures 7–9. It is clear that the efficacies of these

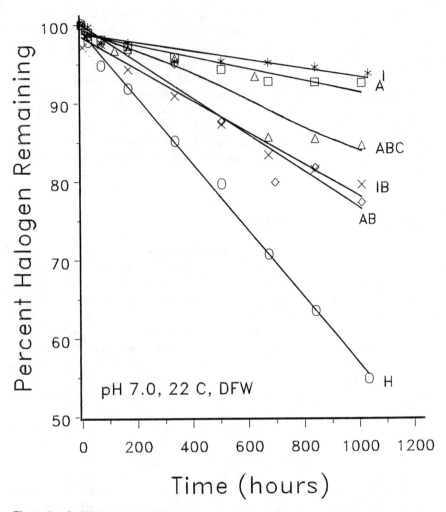

Figure 2.   Stabilities of the disinfectant compounds in demand-free water, pH 7.0, 22°C.

compounds against these two organisms are inversely related to their stabilities. Free chlorine kills these organisms within seconds under these conditions in demand-free water. Compounds A and I provide no measurable free chlorine under these conditions; compound IB provides about 5% free bromine at pH 7.0 in demand-free water at ambient temperature, and compound AB provides about one-half this amount. Thus, organisms that are rapidly inactivated by free halogen would be expected to be more sensitive to inactivation by the combined bromamines IB and AB than by the combined chloramines I and A, with ABC being intermediate. All of the combined halamines tend to function well at high pH (9.5), unlike free chlorine, which

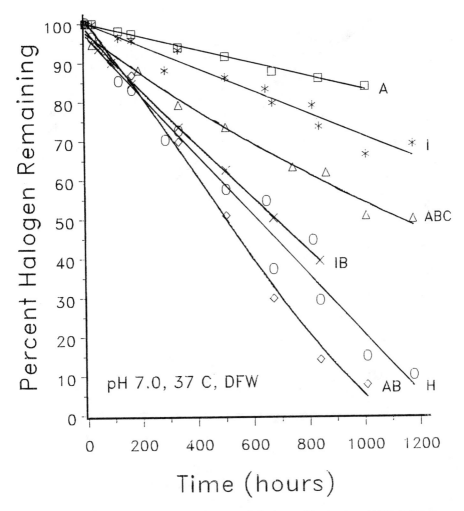

Figure 3. Stabilities of the disinfectant compounds in demand-free water, pH 7.0, 37°C.

is a much better bactericide at low pH. This could be due to increased dissociation of the N-X bond in alkaline solution to provide an increased concentration of free halogen relative to neutral or acidic media. Even though free chlorine is a less efficient bactericide in the hypochlorite form than as hypochlorous acid, free chlorine in any form is more effective against *S. aureus* than is combined halogen. Figure 9 shows that all of the compounds were more effective against *Sh. boydii* than against *S. aureus*. It also shows that compounds I and A were about equal in efficacy against *Shigella*, which was not the case for *Staphylococcus*. Thus, generalizations should not be made concerning relative efficacy against all bacteria; the observations are

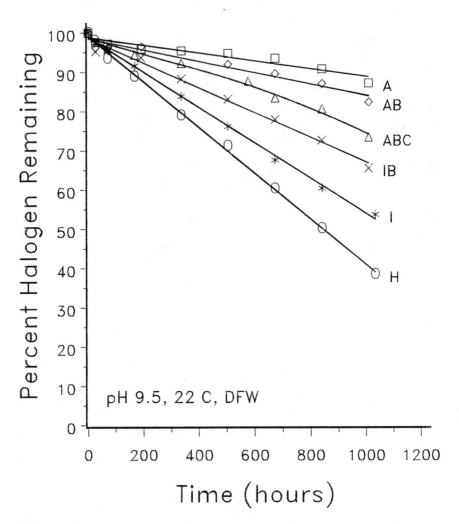

Figure 4.   Stabilities of the disinfectant compounds in demand-free water, pH 9.5, 22°C.

certainly organism-dependent. Tables 1 and 2 summarize the efficacy data for the compounds against the two organisms under a variety of conditions. These efficacies are presented as CT products (concentration in mg/L, time in min necessary for a 6 log inactivation). The lower the CT product, the better was the performance of the disinfectant. For the SDW experiments, the CT products are presented as ranges because the disinfection kinetics were very nonlinear due to the presence of the organic load. Concentrations considered were 1, 2.5, 5, and 10 mg/L total chlorine or molar equivalents in total bromine or total oxidant.

The performance of the various compounds against the protozoa *Enta-*

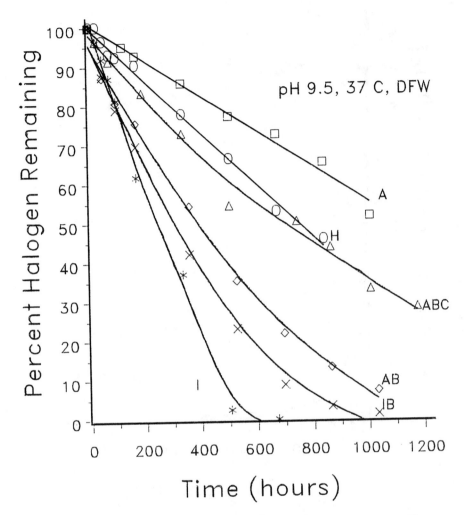

Figure 5. Stabilities of the disinfectant compounds in demand-free water, pH 9.5, 37°C.

moeba invadens and *Giardia lamblia* are indicated as CT products in Table 3. All of the compounds were as effective, or in most cases more effective, than free chlorine against these organisms. Thus, it cannot be generalized that free chlorine is always a more efficient biocide than are combined halamines. Obviously, the efficacy is dictated by the mechanism of action against a given organism. The performance of the new compounds against the protozoa is particularly gratifying because control of *Giardia lamblia* is a serious problem in water treatment.

Other organisms considered in this study were *Legionella pneumophila* and rotavirus. For *Legionella* only compound I and free chlorine were con-

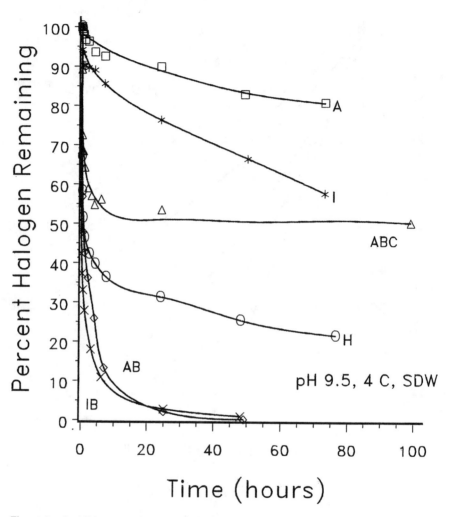

Figure 6.   Stabilities of the disinfectant compounds in synthetic-demand water, pH 9.5, 4°C.

sidered. The CT products for a 4 log inactivation in synthetic-demand water at 4°C and pH 9.5 were <50 mgL^{-1}min and <20 mgL^{-1}min for compound I and free chlorine, respectively. When the much greater stability of compound I than of free chlorine in water containing organic load is considered, it is obvious that I would be effective for applications such as disinfection of closed-cycle cooling water systems and cooling towers for which speed of disinfection is not important. Compound I would maintain the water free of *Legionella* for a long period of time. Free chlorine could not be used for these applications because of its instability and its corrosive properties. The combined halamine compounds are much less corrosive than free halo-

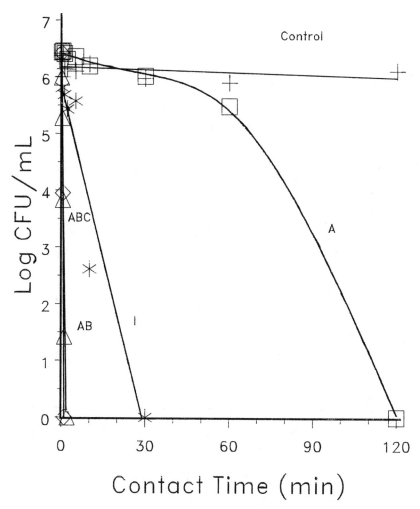

Figure 7. Efficacies of the disinfectant compounds against *S. aureus* in demand-free water, pH 7.0, 22°C at 10 ppm total chlorine or its molar equivalent in total bromine or total oxidant concentration.

gen.[3,17] For rotavirus, the best of the new compounds by far was compound ABC; at 22°C, pH 7.0, demand-free water, the CT product for ABC against this virus was 300 mgL^{-1}min, only a factor of three greater than that for free chlorine. The values for the other new compounds were all $\geq$ 24,000 mgL^{-1}min.

The data for the rechallenge of the various compounds against *S. aureus* at pH 7.0, 22°C, demand-free water at an initial total halogen concentration of 2.5 mg/L total chlorine (or its molar equivalent in total bromine or total oxidant), are presented in Table 4. The contact times given in Table 4

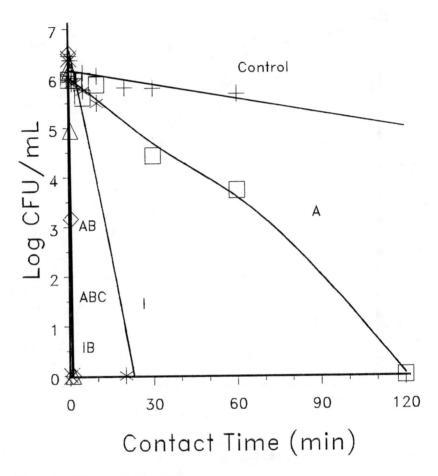

Figure 8. Efficacies of the disinfectant compounds against *S. aureus* in demand-free water, pH 9.5, 22°C at 5 ppm total chlorine or its molar equivalent in total bromine or total oxidant concentration.

correspond to the interval during which the viable cfu/mL declined from any value to zero detectable. From Table 4 it can be seen that IB killed the organisms the most rapidly (1–2 min) through the fourth challenge at 336 hr, but then became completely ineffective after the sixth challenge at 672 hr. Compound AB survived the sixth challenge and became ineffective after the seventh challenge at 1008 hr. Compound ABC killed rapidly through the fourth challenge at 336 hr and then slowly throughout the remainder of the study. The bromine moiety is lost from the molecule between the fourth and fifth challenges, leaving a monochloramine that provides slow but long-term

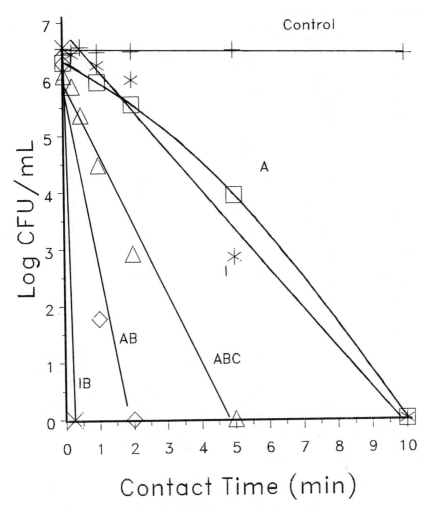

Figure 9. Efficacies of the disinfectant compounds against *Sh. boydii* in demand-free water, pH 7.0, 22°C at 2.5 ppm total chlorine or its molar equivalent in total bromine or total oxidant concentration.

disinfection. Evidence supporting this is shown in Figure 10; the disinfection kinetics of the monochloramine derived from decomposition of ABC (aged ABC) were the same as those for the dichloramine compound A. The monochloramine for this experiment was produced by allowing ABC to contact the synthetic-demand water for ca. 90 hr (see inset of Figure 10) at which time 50% of the total halogen (the chlorine) remained. Compounds I and A killed slowly throughout the whole rechallenge experiment. It would appear that I kills substantially better than does A; however, this is misleading in that the same total chlorine concentration for each (2.5 mg/L) refers to about

Table 1. CT Products for Disinfection of S. aureus

pH	Temp	Demand[a]	Compound	CT[b]
	Test Conditions		**Compound**	**CT[b]**
7.0	22°C	DFW	I	216
7.0	22°C	DFW	IB	3.3
7.0	22°C	DFW	A	1400
7.0	22°C	DFW	AB	9.8
7.0	22°C	DFW	ABC	36.3
7.0	22°C	DFW	Ca(OCl)$_2$	<0.25
4.5	22°C	DFW	I	326
4.5	22°C	DFW	IB	2.1
4.5	22°C	DFW	A	1295
4.5	22°C	DFW	AB	2.4
4.5	22°C	DFW	ABC	13
4.5	22°C	DFW	Ca(OCl)$_2$	<0.25
9.5	22°C	DFW	I	150.8
9.5	22°C	DFW	IB	1.5
9.5	22°C	DFW	A	522.6
9.5	22°C	DFW	AB	3.3
9.5	22°C	DFW	ABC	14.1
9.5	22°C	DFW	Ca(OCl)$_2$	4.7
9.5	4°C	DFW	I	648.5
9.5	4°C	DFW	IB	2.52
9.5	4°C	DFW	A	4355
9.5	4°C	DFW	AB	25.1
9.5	4°C	DFW	ABC	651
9.5	4°C	DFW	Ca(OCl)$_2$	24.1
9.5	4°C	SDW	I	1343–2620[c]
9.5	4°C	SDW	IB	18.5–4940
9.5	4°C	SDW	A	9679
9.5	4°C	SDW	AB	291–6270
9.5	4°C	SDW	ABC	1292–2264
9.5	4°C	SDW	Ca(OCL)$_2$	45.6–16206

[a]DFW = demand-free water; SDW = synthetic-demand water.
[b]CT products are in mgL^{-1}min for complete disinfection of 10^6 cfu/mL.
[c]Intervals indicate nonlinear response.

half as many total molecules for A as for I. In a similar experiment, free chlorine was observed to become ineffective between a third challenge at 72 hr and a fourth one at 96 hr. Thus, the order of biocidal stability of the several disinfectants was observed to be I,A > ABC > AB > IB > Ca(OCl)$_2$. It would appear that compound ABC represents the best compromise disinfectant in that the bromine moiety provides rapid initial kill, while the chlorine moiety provides slow, long-term kill of the organisms.

The structural feature present in all of the new compounds which renders them exceptionally stable in aqueous solution is the two methyl groups adjacent to the N-X functional groups on the rings. There are at least three possible explanations for this. First, the methyl groups being electron-donating substituents destabilize developing negative charge on nitrogen as positive halogen leaves, thus stabilizing the N-X bond. Second, the presence of two methyl groups on the carbon adjacent to the N-X bond prevents a

Table 2. CT Products for Disinfection of *Sh. boydii*

| | Test Conditions | | | |
pH	Temp	Demand[a]	Compound	CT[b]
7.0	22°C	DFW	I	23.4
7.0	22°C	DFW	IB	1.3
7.0	22°C	DFW	A	26
7.0	22°C	DFW	AB	9.7
7.0	22°C	DFW	ABC	15.1
7.0	22°C	DFW	$Ca(OCl)_2$	<0.25
9.5	4°C	DFW	I	ND[c]
9.5	4°C	DFW	IB	<2.5
9.5	4°C	DFW	A	104–153[d]
9.5	4°C	DFW	AB	24.8
9.5	4°C	DFW	ABC	13.7
9.5	4°C	DFW	$Ca(OCl)_2$	<2.5
9.5	4°C	SDW	I	11.91
9.5	4°C	SDW	IB	26.2–854
9.5	4°C	SDW	A	262–280
9.5	4°C	SDW	AB	148–314
9.5	4°C	SDW	ABC	149
9.5	4°C	SDW	$Ca(OCl)_2$	<2.5

[a]DFW = demand-free water; SDW = synthetic-demand water.
[b]CT products are in $mgL^{-1}min$ for complete disinfection of 10^6 mL.
[c]ND = no determination.
[d]Intervals indicate nonlinear response.

Table 3. CT Products for Disinfection of Protozoa in Demand-Free Water, pH 7.0, 22°C

Organism	Compound	CT[a]
Entamoeba invadens	I	1
Trophozoites	IB	4–10[b]
	A	10
	AB	1
	ABC	4
	$Ca(OCl)_2$	10
Giardia lamblia	I	1
Trophozoites	IB	4
	A	4–10
	AB	1
	ABC	2
	$Ca(OCl)_2$	4
Giardia lamblia	I	10
Cysts	$Ca(OCl)_2$	20

[a]CT products are in $mgL^{-1}min$ for complete disinfection (6 log).
[b]Intervals indicate nonlinear response.

Table 4.   Disinfection Intervals (Min) for Rechallenges of N-Halamine Compounds[a]

Rechallenge Time (hr)	I	IB	A	AB	ABC
0	60(3)– 120(0)	1(4)– 2(0)	120(3)– 1440(0)	1(3)– 2(0)	1(3)– 2(0)
48	60(3)– 120(0)	1(4)– 2(0)	180(1)– 1440(0)	2(1)– 5(0)	1(2)– 2(0)
168	60(1)– 120(0)	1(4)– 2(0)	108(2)– 1440(0)	1(1)– 2(0)	1(1)– 2(0)
336	60(1)– 120(0)	1(4)– 2(0)	165(1)– 1440(0)	2(2)– 5(0)	5(1)– 10(0)
504	30(3)– 60(0)	120(1)– 1440(0)	120(2)– 1440(0)	10(3)– 30(0)	60(4)– 1440(0)
672	60(2)– 120(0)	1440(4)	240(1)– 360(0)	30(3)– 60(0)	420(3)– 1440(0)
1008	60(1)– 120(0)	1440(4)	120(2)– 240(0)	1440(4)	360(4)– 1440(0)
1704	60(3)– 120(0)	1440(4)	300(1)– 1440(0)	1440(4)	1440(1)
2856	60(3)– 120(0)	1440(4)	360(2)– 1440(0)	1440(4)	1440(2)

[a]Rechallenges by fresh inocula of *Staphylococcus aureus* at 10^6 cfu/mL. Initial halogen concentration was 7.05×10^5 M (equivalent to 2.5 mg/L Cl[+]). Numbers in parentheses are the subjective score of colony density at the indicated contact time (0 = no detectable survivors, 1 = less than 50 cfu/25 μL, 2 = greater than 50 cfu/25 μL but distinct colonies, 3 = too numerous to count but not confluent, and 4 = confluent growth). The water was DFW buffered at pH 7.0 and held at 22°C.

possible dehydrohalogenation reaction. Third, the bulky methyl groups may sterically hinder the solvolysis of the N-X moiety. We are currently performing experiments designed to distinguish among these possibilities.

CONCLUSIONS

It can be concluded from this work that the N-halooxazolidinones and N,N'-dihaloimidazolidinones represent useful new classes of disinfectants for water applications. The great stability of these compounds renders them particularly useful for disinfection applications for which contact time is not of primary importance. Such applications include, but are not limited to, cooling towers, air conditioning systems, swimming pools (particularly winterization), and hot tubs. We would recommend compound ABC (1-bromo-

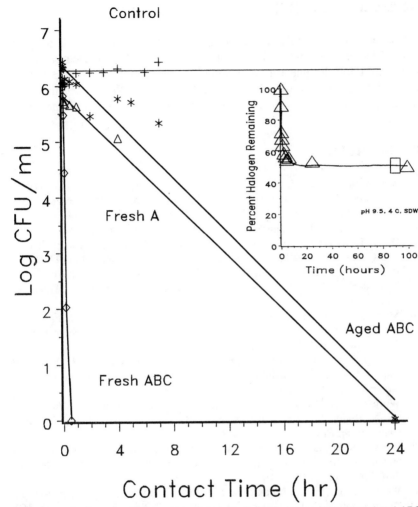

Figure 10. Disinfection efficacies for compounds A, ABC, and partially dissociated ABC in synthetic-demand water at a total chlorine concentration of 5 ppm or its molar equivalent in total bromine or total oxidant; inset shows that the aged ABC was dissociated to the extent of 50 percent.

3-chloro-4,4,5,5-tetramethyl-2-imidazolidinone) as the best for most applications because of its rapid initial disinfection from the bromine moiety and slower, long-term disinfection from the chlorine moiety. It should be stressed that much toxicity testing employing laboratory animals remains to be performed, but that the results of initial testing on poultry were favorable.[4]

ACKNOWLEDGMENTS

The authors gratefully acknowledge the generous support of the U.S. Army Medical Research and Development Command at Ft. Detrick and the U.S. Air Force Engineering and Services Center at Tyndall AFB through contract DAMD 17–82-C-2257. We also thank the Water Resources Research Institute at Auburn University for the administration of the project.

REFERENCES

1. Kaminski, J. J., M. H. Huycke, S. H. Selk, N. Bodor, and T. Higuchi. "N-Halo Derivatives V. Comparative Antimicrobial Activity of Soft N-Chloramine Systems," *J. Pharm. Sci.* 65(12):1737–1742 (1976).
2. Kosugi, M., J. J. Kaminski, S. H. Selk, I. H. Pitman, N. Bodor, and T. Higuchi. "N-Halo Derivatives VI. Microbiological and Chemical Evaluations of 3-Chloro-2-oxazolidinones," *J. Pharm. Sci.* 65(12):1743–1746 (1976).
3. Burkett, H. D., J. H. Faison, H. H. Kohl, W. B. Wheatley, S. D. Worley, and N. Bodor. "A Novel Chlorine Compound for Water Disinfection," *Water Res. Bull.* 17(5):874–879 (1981).
4. Mora, E. C., H. H. Kohl, W. B. Wheatley, S. D. Worley, J. H. Faison, H. D. Burkett, and N. Bodor. "Properties of a New Chloramine Disinfectant and Detoxicant," *Poul. Sci.* 61:1968–1971 (1982).
5. Worley, S. D., W. B. Wheatley, H. H. Kohl, J. A. Van Hoose, H. D. Burkett, and N. Bodor. "The Stability in Water of a New Chloramine Disinfectant," *Water Res. Bull.* 19(1):97–100 (1983).
6. Worley, S. D., W. B. Wheatley, H. H. Kohl, H. D. Burkett, J. A. Van Hoose, and N. Bodor. "A New Water Disinfectant; A Comparative Study," *Ind. Eng. Chem. Prod. Res. Dev.* 22(4):716–718 (1983).
7. Worley, S. D., and H. D. Burkett. "The Stability of a New Chloramine Disinfectant as a Function of pH, Temperature, and Water Quality," *Water Res. Bull.* 20(3):365–368 (1984).
8. Worley, S. D., H. D. Burkett, and J. F. Price. "The Tendency of a New Water Disinfectant to Produce Toxic Trihalomethanes," *Water Res. Bull.* 20(3):369–371 (1984).
9. Worley, S. D., W. B. Wheatley, H. H. Kohl, H. D. Burkett, J. H. Faison, J. A. Van Hoose, and N. Bodor. "A Novel Bactericidal Agent for Treatment of Water," in *Water Chlorination: Environmental Impact and Health Effects, Vol. 4*, R. L. Jolley, W. A. Brungs, J. A. Cotruvo, R. B. Cumming, J. S. Mattice, and V. A. Jacobs, Eds. (Ann Arbor, MI: Ann Arbor Science Publishers, Inc., 1983), 1105–1113.
10. Worley, S. D., D. E. Williams, H. D. Burkett, L. J. Swango, C. M. Hendrix, and M. H. Attleberger. "Comparisons of a New N-Chloramine Compound with Free Chlorine as Disinfectants for Water," in *Prog. in Chem. Disinfect., Vol. 2*, G. E. Janauer, Ed. (Binghamton: SUNY, 1984), 45–60.

11. Worley, S. D., D. E. Williams, H. D. Burkett, S. B. Barnela, and L. J. Swango. "Potential New Water Disinfectants," in *Water Chlorination: Environmental Impact and Health Effects, Vol. 5,* R. L. Jolley, R. J. Bull, W. P. Davis, S. Katz, M. H. Roberts, Jr., and V. A. Jacobs, Eds. (Chelsea, MI: Lewis Publishers, Inc., 1985), 1269–1283.

12. Williams, D. E., S. D. Worley, W. B. Wheatley, and L. J. Swango. "Bactericidal Properties of a New Water Disinfectant," *Appl. Environ. Microbiol.* 49(3):637–643 (1985).

13. Ahmed, H., J. L. Aull, D. E. Williams, and S. D. Worley. "Inactivation of Thymidylate Synthase by Chlorine Disinfectants," *Int. J. Biochem.* 18(3):245–250 (1986).

14. Worley, S. D., D. E. Williams, S. B. Barnela, E. D. Elder, L. J. Swango, and L. Kong. "New Halamine Water Disinfectants," in *Prog. in Chem. Disinfect., Vol. 3,* G. E. Janauer, Ed. (Binghamton: SUNY, 1986), 61–79.

15. Kaminski, J. J., and N. Bodor. "3-Bromo-4,4-dimethyl-2-oxazolidinone," *Tetrahedron* 32:1097–1099 (1976).

16. Barnela, S. B., S. D. Worley, and D. E. Williams. "Syntheses and Antibacterial Activity of New N-Halamine Compounds," *J. Pharm. Sci.* 76(3):245–247 (1987).

17. Worley, S. D., L. J. Swango, M. H. Attleberger, C. M. Hendrix, H. H. Kohl, W. B. Wheatley, D. E. Williams, H. D. Burkett, D. Geiger, and T. Clark. "New Disinfection Agents for Water," Report ESL-TR-83–68, AD A149537, Engineering & Sciences Laboratory, Air Force Engineering & Service Center, Tyndall AFB, 1984, 74 pp.

INDEX

Acetic acid. (*See* Aliphatic acids)

Acetone. (*See* Carbonyl compounds)

Activated carbon. (*See* Granular activated carbon)

Advanced oxidation processes, 185

Aeromonas. (*See* Bacteria)

Aldehydes. *(See also* Carbonyl compounds)
  biodegradability of, 159
  formation by ozonolysis, 156, 160, 190, 203–209
  reaction with hydrogen peroxide, 188–189, 192
  reactions with chloramine, 157–158
  removal by filtration, 160

Aliphatic acids
  fatty, effects of ozone on, 207–209
  formation by ozonolysis, 157, 190
  products of humic acid chlorination, 111–114
  reaction with ozone—UV, 190, 195–196

Alkali, effect on *Legionella*, 76–77

Ames test. (*See* Mutagens and carcinogens)

Amino acids
  chlorination of, 104–105, 133–138
  in stomach fluid, 143

Ammonium nitrogen, 258–259

Aspartic acid. (*See* Amino acids)

Assimilable organic carbon, 259–261

Bacteria in water systems
  Aeromonas, 10–11
  autotrophic, 260
  biofilms, 259–261
  chlorine effect on attached, 248
  chlorine effect on *Legionella*, 68–70

coliforms, 61–62, 68, 244–245, 247, 249–250
  heterotrophic, 245–249
  Klebsiella, 247
  Legionella, 67–80, 273, 277–278
  nitrifying, 260–261
  oligotrophic, 259–261
  Pseudomonas, 61–62, 69
  Salmonella, 249–250
  Shigella, 249, 273–276, 281, 283
  Staphylococcus, 273–276, 279–280
  Yersinia, 249–251

Behaviorism, in risk perception, 22

Benzenecarboxylic acids, from chlorination of humic substances, 98

Benzofurans. (*See* Chlorinated benzofurans)

Benzoquinones. (*See* Quinones)

Bicarbonate, effect on ozone in water, 154

Biologically stable and instable water, 257–261

Bromide, oxidation by ozone, 155–156

3-Bromo-4,4-dimethyl-2-oxazolidinone. (*See* Halamines, disinfection)

Carbonate. (*See* Bicarbonate)

Carbonyl compounds
  derivatization with 2,4-dinitrophenylhydrazine, 186–191
  formation during ozonolysis in water, 157, 190
  reaction with hydrogen peroxide, 187, 192

Carcinogenic effects
  animal tests, 7
  complex organic mixtures, 8
  epidemiological studies, 6, 8

Chloral, 97, 115, 134–138

289